Essential Guide to Economics Made Simple

James D. Reed

All rights reserved. Copyright © 2023 James D. Reed

Funny helpful tips:

Engage in scenario planning; it prepares you for various market conditions.

Stay informed about health trends; however, always consult professionals before making significant changes to your regimen.

Essential Guide to Economics Made Simple : Master the Principles of Economics with This User-Friendly Handbook for Beginners

Life advices:

Your voice is a powerful instrument; use it to echo love, hope, and inspiration.

Stay proactive in injury prevention; warm-ups, cool-downs, and proper techniques are essential.

Introduction

Welcome to this book, a comprehensive guide that will take you on a journey through the fascinating world of economics. Whether you're a student, a professional, or simply curious about how economies function, this book will provide you with a solid foundation and a clear understanding of key economic principles and concepts.

In the opening chapters, we will explore the fundamental question of what economics is all about. You'll gain insights into the basic principles that govern economic behavior and decision-making. From there, we'll delve into the reasons why we engage in trade and the benefits it brings to individuals, businesses, and nations.

One of the central aspects of economics is the study of money and its role in our society. We'll take a closer look at the story of money, its historical development, and how it functions as a medium of exchange, unit of account, and store of value. We'll also explore the workings of banks and the role they play in facilitating economic transactions and providing financial services.

Understanding supply and demand is crucial to comprehending how markets operate. We'll delve into the forces that shape supply and demand, and how they interact to determine prices and quantities in competitive markets. Additionally, we'll examine imperfectly competitive markets and the impact of market power on pricing and competition.

The role of government in the marketplace is a vital aspect of economics. We'll explore how governments intervene to regulate markets, promote fairness, and address market failures. We'll also analyze the role of financial markets and their significance in mobilizing capital, facilitating investment, and supporting economic growth.

As economies become increasingly interconnected, foreign exchange and the balance of payments play a critical role. We'll examine how international trade and exchange rates impact economies and nations' economic well-being. Furthermore, we'll explore the circular flow of economic activity, which illustrates the interdependence of households, businesses, and the government in an economy.

Measuring economic performance is essential to understanding the health of an economy. We'll dive into key indicators such as the Gross Domestic Product (GDP) and explore how it is calculated and interpreted. We'll also analyze the factors that contribute to economic growth and the consequences of inflation on individuals and societies.

Macroeconomic equilibrium and the role of the Federal Reserve in monetary policy are vital topics we'll explore. You'll learn about the tools and strategies employed by central banks to manage interest rates, control inflation, and stabilize economies. We'll also discuss the controversies and debates surrounding economic policies, including the concept of "voodoo economics."

In recent years, the global economy faced significant challenges, including the Great Recession. We'll examine the causes and consequences of this economic downturn and explore the lessons learned from the experience. Furthermore, we'll discuss the relationship between the environment and the economy, highlighting the importance of sustainable practices and addressing environmental concerns.

By the end of this book, you'll have a comprehensive understanding of the fundamental principles of economics, their applications, and their impact on individuals, societies, and the global economy. Whether you're interested in personal finance, business decision-making, or public policy, the insights gained from this book will empower you to make informed choices and contribute to economic discussions and debates.

So, let's embark on this enlightening journey into the world of economics and explore the fascinating dynamics that shape our lives and the world we live in.

Contents

CHAPTER 1: What Is Economics?..1
Economics..1
Microeconomics...2
Macroeconomics..3
Scarcity..3
Trade-Offs and Opportunity Cost...5
Marginal Analysis..6
Assumptions in Economics..7
CHAPTER 2: Why Do We Trade?..10
History of Trade...10
Mercantilism..11
Free Trade...11
Comparative Advantage: A Not So Obvious Theory.......................................12
Absolute Advantage..12
Comparative Advantage..12
The Gains from Trade..15
International Trade..16
The Case for Free International Trade..16
The Case Against International Trade..18
Other Bases of Trade..18
A Look at Trade Barriers...20
Tariffs..21
Quotas...22
Embargoes..22

CHAPTER 3: That's Not the Way We Do Things! 24
Different Economic Systems 24
Traditional Economies 25
Command Economies 26
Market Economies 28
Capitalism Versus Socialism 30
Adam Smith, Karl Marx, and Human Nature 32
CHAPTER 4: The Story of Money 34
Barter 34
The Development of Money 35
Functions and Characteristics of Money 35
Money Throughout History 36
The Gold Standard 38
Welcome to the Matrix 39
M1 and M2 41
The Time Value of Money and Interest Rates 42
CHAPTER 5: Banks 45
The Origin of Banking 45
The Function of Banks 46
Balance Sheets 48
How Banks Create Money 49
Banks as a System 51
Bank Runs 52
Bank Regulation and Deregulation 53
Bank Regulations Since the 1900s 54
Bank Deregulation 55
CHAPTER 6: Supply and Demand 57

Markets	57
Consumer Behavior	59
Demand	60
Reasons for the Law of Demand	61
Elasticity of Demand	62
Changes in Demand	63
Supply	66
Elasticity of Supply	66
Changes in Supply	67
A Price Is Born	69
Finding Equilibrium	71
Changes in Demand and Supply	72
Markets Talk	74
CHAPTER 7: Competitive Markets	76
Conditions	76
Accounting Versus Economics	77
The Production Function	79
The Stages of Production Function	79
Cost, Cost, Cost	81
Perfect Competition in the Short Run	82
Perfect Competition in the Long Run	83
From Competition to Imperfect Competition: AContinuum	84
Monopolistic Competition	85
CHAPTER 8: Imperfectly Competitive Markets	88
Oligopoly	88
Collusion and Cartels	90
Forming Cartels	91

Game Theory	92
Game Theory in Business	94
Pricing Behaviors	96
Monopolies	97
Price Discrimination	98
The Good, the Bad, and the Ugly	99
The Good	99
The Bad?	101
The Ugly	102
CHAPTER 9: Government in the Marketplace	103
Price Ceilings	103
Price Floors	105
Taxes and Subsidies	107
Market Failures	108
Public Goods	108
Positive Externality	109
Negative Externality	110
Black Markets	111
CHAPTER 10: Financial Markets	115
Loanable Funds Theory	115
Liquidity Preference	117
The Money Market	118
Bond Market	120
Types of Bonds	121
Bond Risks	123
Interest Rates Revisited	124
Stock Market	124

So What's the Point? 126
CHAPTER 11: Foreign Exchange and the Balance of Payments 128
Exchange Rates 128
Taste and Preference 130
Interest Rates and Inflation 130
Relative Income and Speculation 131
The Law of One Price and Purchasing Power Parity 133
Net Exports 134
Net Foreign Factor Income 135
Net Transfers 135
Net Foreign Investment 136
Official Reserves 138
To Fix or Float? That Is the Question 139
CHAPTER 12: The Circular Flow of Economic Activity 142
The Private Sector 142
The Public Sector 144
The Foreign Sector 145
The Financial Sector 146
The Financial Markets 147
CHAPTER 13: Keeping Score: The Gross Domestic Product 150
Wealth and Income 150
A Look Back at Circular Flow 151
Counted or Not? 152
The Spending Approach 154
Personal Consumption Expenditure 154
Gross Private Investment 156
Government Spending 158

Net Exports 159
Income Approach to GDP 160
Nominal Versus Real 160
Real GDP Changes and the Business Cycle 161
What GDP Doesn't Tell Us 163
CHAPTER 14: Where Did My Job Go? 166
What Unemployment Is and What It Is Not 166
Measuring Unemployment 167
Falling Through the Cracks 169
Types of Unemployment 170
Frictional Unemployment 170
Structural Unemployment 171
Cyclical Unemployment 172
Full Employment 173
Why Unemployment Is Bad 174
Trends and Demographics 176
CHAPTER 15: Inflation 178
What Is Inflation? 178
Measuring Inflation 179
Types of Inflation 181
Demand-Pull Inflation 182
Cost-Push Inflation 183
Who Gains and Who Loses from Inflation? 184
Benefiting from Inflation 184
Losing with Inflation 185
Disinflation 187
The Role of Expectations 188

Deflation 189
CHAPTER 16: Putting It All Together: MacroeconomicEquilibrium........................192
Aggregate Demand..192
Aggregate Supply..193
The Short Run..194
The Long Run 194
Changes in SRAS 195
Changes in LRAS 195
Macroeconomic Equilibrium 196
The Classical View..197
The Keynesian View and Fiscal Policy 199
The Inflation-Unemployment Trade-Off 202
CHAPTER 17: The Federal Reserve and Monetary Policy 206
History of Banking Revisited 206
The Federal Reserve System 208
The FOMC 209
Monetary Policy..211
Reserve Requirement 212
Discount Rate..213
Open Market Operations..214
When Policies Collide: Fiscal/Monetary Mixer..216
Monetary Policy in the Short Run and the Long Run 217
CHAPTER 18: Voodoo Economics 219
Stagflation 219
Thinking Outside the Box..220
Supply-Side Economics 222
Reagan, Cheney, and Laffer...224

Challenges to Supply-Side Economics	224
A Complete Toolbox	227
CHAPTER 19: Economic Growth	230
What Growth Means	230
Why Grow?	231
Human Capital	233
Physical Capital	234
Research and Development	236
The Rule of Law	236
How Economic Policy Affects Growth	237
The Downsides to Economic Growth	240
CHAPTER 20: The Great Recession of 2007–20??	242
The Run-Up to the Meltdown	242
Securitization	244
Risk Management and Credit Default Swaps	245
The Collapse of Investment Banking	247
A Coordinated Policy Response	248
The Fed	249
Quantitative Easing	249
Government Gets Inventive	250
Theory Versus Practice	251
CHAPTER 21: The Environment and the Economy	255
Is Growth Sustainable?	255
Renewable Resources	256
Nonrenewable Resources	256
Tree Huggers and Money Grubbers — Can't We All Get Along?	257
Tragedy of the Commons	258

Human Population 261
The Role of Incentives 261
The Right to Pollute 262
Per Unit Taxes 263
Pollution Permits 264
The Coase Theorem 265

CHAPTER 1: What Is Economics?

You open the door to your fridge and gaze at the food inside and declare, "There's nothing to eat in this house." Later, you walk into a closet full of clothes and then think, "I have nothing to wear." You are faced with scarcity. There never is enough of what you need or want. The fact is, there is plenty to eat and there are many clothes to wear. You chose to ignore the options you faced then and there, but eventually you know you will relent and eat the apple next to the shriveled grapes at the bottom of the bin, and then put on the shirt and pants you hate. You are a creature of economics. Given scarcity, you look at the choices you face, evaluate, and then choose.

Economics

Economics is the study of how individuals, institutions, and society choose to deal with the condition of scarcity. It is fascinating to see how people react to scarcity. Some create complex plans and systems to make sure that everyone gets their fair share of scarce resources. Others make things up as they go along. Everybody practices economics on a daily basis. From a single individual to the largest society on earth, people are constantly engaged in the struggle to survive, make ends meet, and even thrive given the relative scarcity they face.

FACT

Economics has been around a long time, though it has not always been known by that name. Philosophers studied scarcity and choice long before the field was so named. The father of modern economics, Adam Smith, was considered a moral philosopher, not an economist.

The people who study these choices are economists. The field of economics is huge because people have an immense range of choices. Some economists study the decision-making of individuals and institutions; others study how nations handle scarcity. Economists develop theories to explain the behavior of whatever it is they are studying. Some of these theories are then tested against real-world data, and sometimes these theories are put into practice without ever being tested. Economists work for universities, financial institutions, major corporations, and governments.

Microeconomics

The field of **microeconomics** focuses its attention on the decision-making of individuals and businesses. Microeconomics is primarily concerned with markets for goods, services, and resources. Markets are central to understanding microeconomics. Whenever and wherever buyers and sellers come together to exchange resources, goods, or services, a market is created, and the behavior of these markets is of particular interest to economists. Are they functioning efficiently? Do participants have access to adequate information? Who and how many participate in the market? How do the decisions made in one market impact the decisions in a related market?

Macroeconomics

Macroeconomics is the study of how entire nations deal with scarcity. Macroeconomists analyze the systems nations create or allow for the allocation of goods and services. The questions they ask are varied and of great interest to individuals and policymakers alike. How do you measure the economy? Why does unemployment exist? How do changes in the amount of money affect the entire economy? What impact does government spending or tax policy have on the economy? How can you make the economy grow?

Scarcity

Without scarcity there would be no need for the study of economics. For that matter, if scarcity did not exist, there would be no need for this book. You are not that lucky. **Scarcity** is the universal condition that exists because there is not enough time, money, or stuff to satisfy everyone's needs or wants. The stuff that everyone wants is made from resources. In an effort to make economics sound more "economic-y," resources are referred to as the factors of production. The factors of production include land, labor, capital, and entrepreneurship.

QUESTION

Is there really scarcity in America, a land of plenty?

Scarcity exists for everyone. From rich to poor, all face the condition. Scarcity in America looks different than scarcity in Somalia to be sure. Here, there is plenty of food and clean water, but in Somalia both are lacking. Scarcity isn't just a function of limited resources but also of unlimited wants, and that is something both America and Somalia share.

Land is inclusive of all natural resources and not just some random piece of property. Trees, mineral deposits, fish in the ocean, ground water, and plain old land are all included. Land can be divided into renewable and nonrenewable natural resources. Renewable resources, like pine trees and chickens, are easily replenished. Nonrenewable resources, like oil and Atlantic cod, are difficult to replenish. The payment for land is referred to as **rent**.

Labor refers to people with their skills and abilities. Labor is divided into unskilled, skilled, and professional. Unskilled labor refers to people without formal training that are paid wages to do repetitive tasks like make hamburgers or perform assembly line production. Skilled labor refers to people paid wages for what they know and what they can do. Welders, electricians, plumbers, mechanics, and carpenters are examples of skilled labor. Professional labor is paid wages for what they know. Doctors, lawyers, engineers, scientists, and even teachers are included in this category.

ESSENTIAL

Economists describe getting the right resources to the right people as allocation. Allocative efficiency occurs when

MB = MC → Best for society

marginal benefit equals marginal cost. When this condition is met the greatest benefit accrues to society.

Capital in economics does not refer to money but to all of the tools, factories, and equipment used in the production process. Capital is the product of investment. Stop. Isn't that confusing? Up until now you have probably lived a happy life thinking that capital was money and that investing is what you do in the stock market. Well, sorry. Capital is physical stuff used to make other stuff and investment is the money spent on buying that stuff. To make capital, you have to have capital. Because capital is always purchased with borrowed money it incurs an interest payment.

Unlike labor, when one ends up working for someone else, **entrepreneurship** actually creates businesses by combining land, labor, and capital in new ways to provide a good or service. Entrepreneurs are unique individuals who are willing to take great risks. These people are willing to risk their wealth in order to earn profits. Entrepreneurs include such well-known people as Bill Gates, Richard Branson, Oprah Winfrey, Ray Kroc, Mary Kay Ash, and that little girl down the street who sets up a lemonade stand when the weather gets hot.

Trade-Offs and Opportunity Cost

Whenever you use a factor of production, a cost is going to be incurred. Why? The factors of production are limited, not limitless. As a result, whenever you choose to use land, labor, capital, or entrepreneurship for one purpose, you lose the ability to use it for another. Take a resource like labor — your labor. Say that you can spend an hour writing a book, teaching a class, or weaving a hammock. The choices you face are called trade-offs. Assume you choose to weave a hammock. You can neither teach a class nor write a book in that hour of time. If writing a book is your next best alternative, then economists would say that the opportunity cost of spending an hour weaving a hammock is the hour you could have spent writing a book. **Opportunity cost** is the next best alternative use of a resource.

FACT

Opportunity cost is sometimes referred to as implicit cost. For any productive activity there are explicit costs like labor, raw materials, and overhead, which are easily calculated, and there are the implicit costs, which are more difficult to assess.

For example, suppose it's a beautiful Friday morning and you think to yourself, "I could go to work like I'm supposed to, I could stay home and sleep away the day, or I could fly to Cozumel and hang out on the beach and do some scuba diving." Assume that you choose to take the trip to Cozumel, but going to work was your next best alternative. What was the cost of your trip? You paid for the taxi to the airport, the plane ticket, an all-inclusive hotel package, and a dive on Palancar Reef. Was that your only cost? No. You also sacrificed the money you could have made working. Opportunity cost is a bummer. Make sure to always count it when making a decision.

Marginal Analysis

Economists like to think of people as little computers who always count the benefit of their decisions versus the cost of those decisions. Because you usually make decisions one at a time, economists refer to the benefit of a decision as **marginal benefit**. Marginal benefit can be measured in dollars or utils, whichever you prefer. **Utils** are the amount of utility or happiness you get from doing something. They can be converted into dollars easily.

Say that you like to swim laps in the pool for an hour. How many utils do you receive from swimming laps? How much would you have to be paid to not swim laps? If your friend were to keep offering you ever increasing amounts of money to not swim in the pool, then it is probably safe to assume that the dollar amount you accept to not swim in the pool is at least equal to the amount of happiness or utility you would have received had you taken a swim. If it takes $20 to keep you from swimming, then you value swimming no more than $20. Swimming is worth 20 utils to you.

Marginal cost is a related concept. Marginal cost is simply what it costs to either produce or consume one extra unit of whatever it is you are producing or consuming. Go back to the swimming example. Assume that swimming in the pool has a marginal cost of $5. If you earn 20 utils from swimming, would you pay $5 to earn $20 worth of benefit? Of course you would. Now assume that swimming in the pool costs $20.01. Would you spend $20.01 to earn $20 worth of benefit? Probably not. Economists conclude that you will swim as long as the marginal benefit exceeds or equals the marginal cost. For you that means you will swim as long as the marginal cost is less than or equal to $20. If the marginal benefit outweighs the marginal cost, you would probably do it. If the marginal benefit is less than the marginal cost, you probably would not do it. If the marginal benefit equals the marginal cost, it means you are indifferent.

Assumptions in Economics

Whenever economists make an argument such as: "If income taxes fall, then consumption increases," it should be understood as: "If income taxes fall *and nothing else changes*, then consumption increases." Did you catch the difference between the two statements? *And nothing else changes* is also referred to as the **ceteris paribus assumption**. Loosely translated, *ceteris paribus* means "to hold all other things constant." So as you continue reading the book, remember that all statements about cause and effect relationships are made with the *ceteris paribus* assumption.

Another assumption made by economists, and a big one at that, is that people behave rationally. Economists assume that people's choices are made with all available information taken into account as well as the costs and benefits of the choice. Furthermore, economists assume that the choices make sense. The assumption that people behave rationally is subject to debate among different schools of economic thought, but for most economic decisions it is a useful assumption.

FACT

> The twentieth-century economist, John Maynard Keynes, suggested that people do not always behave rationally. He argued that people are motivated at times by fear or hubris in what he termed the "animal spirits."

The last assumption made by economists is that people are self-interested. First and foremost, people think of themselves whenever it comes time to make a decision. Pure altruism is not possible in economics. Economists cynically assume that human behavior is motivated by self-interest. For example, a grenade is thrown into a trench with a platoon of soldiers and one soldier sacrifices his life by jumping on top of the grenade thus saving the others. To economists, this soldier instantly calculated the marginal benefit and marginal cost of the decision, determined that the marginal benefit of saving his fellow soldiers outweighed the marginal cost of his life, and jumped on the grenade as an act of utility, maximizing self-interest. He saved his friends in order to maximize *his* utility as a soldier.

The assumptions economists make are subject to criticism and debate. Many critics believe that the field has a tendency to be too abstract and theoretical to have any real-world value. The failure of most economists to predict the most recent economic downturn seems to support the view that economics ignores human psychology at its own peril.

Economics is at turning point as a field of study, and the assumptions that economists hold dear need to be carefully examined. Instead of being tidy, abstract, and mathematical like physics, economics must become a little more messy, complex, and organic, like biology.

CHAPTER 2: Why Do We Trade?

Which leads to a higher standard of living, dependence on others or complete self-sufficiency? Before you answer the question, consider which approach better represents your life. Do you have a job and pay all your own bills, or do you live at home with your parents while they foot the bill for your upkeep? If you are on your own, you most likely consider yourself to be self-sufficient. If you are still living at home with your parents, you probably consider yourself to be somewhat dependent. The truth, however, is, whether you are on your own or living at home, you are highly dependent on others for the food you eat, the clothes you wear, and the roof over your head. In order for you to get what you need and want and enjoy a higher standard of living, then you must trade with others.

History of Trade

For as long as there have been people, there has been **trade**. At first, trade was a simple matter. For example, people in a family exchanged food with their neighbors. Over time, trade expanded as people were exposed to new goods from faraway places and developed a taste for them. As tribes became kingdoms and kingdoms became empires, trade grew in importance. This growth in trade led to the emergence of the influential merchant class. These merchants braved hardships in search of profit, and their activities helped to form the modern world. Although the scale of trade has grown incredibly throughout history, what has not changed is that trade always occurs between individuals.

Mercantilism

One early theory of trade was **mercantilism**, which dominated seventeenth- and eighteenth-century European trade policy. Mercantilism is founded on the idea that a country and, therefore, individuals, are better off if the value of a country's exports are greater than the value of its imports. Under mercantilism, the more gold a country amassed, the wealthier it became. As a result, countries competed to import cheap natural resources and then convert them into more expensive manufactured goods for export. It is easy to see why the countries of Europe were eager to compete against each other in order to colonize and exploit the newly discovered and resource-rich Americas.

Mercantilism had an obvious flaw. If a country is always trying to export more than it imports and everyone else is playing the same game, then someone is going to lose. In order to maintain a country's export advantage, governments enacted many laws and taxes that distorted the flow of goods without necessarily making the people better off. In the end, mercantilism created a win/lose condition that harmed more than it helped.

Free Trade

The insights of eighteenth-century Scottish thinker Adam Smith were influential in bringing an end to mercantilism. He and others at the time saw governments' mercantilist policies as misguided and prone to influence by special interests. He argued in *Wealth of Nations* that if a country specialized in what it produced best and freely traded those products, then society would be better off. Adam Smith saw wealth as being the sum total of all that the people of a nation produced. In his view, **free trade** led to greater wealth, even if it meant that sometimes you imported manufactured goods from people in other countries.

Comparative Advantage: A Not So Obvious Theory

One argument used in support of the idea of free trade is the theory of comparative advantage. Whereas Adam Smith had argued for a country to specialize in what it does best and then trade with others, another influential thinker, David Ricardo, argued that it is better for a country to specialize in what it produces at the lowest opportunity cost, and then trade for whatever else it needs. These two concepts are referred to as **absolute advantage** and **comparative advantage**.

Absolute Advantage

An absolute advantage exists if you can produce more of a good or service than someone else, or if you can produce that good or service faster than someone else. An absolute advantage implies that you are more efficient, that is, able to produce more with the same amount of resources. For example, Art can write one hit song per hour, whereas Paul can write two hit songs per hour. Thus, Paul has an absolute advantage in songwriting.

Comparative Advantage

A comparative advantage exists if you can produce a good at a lower opportunity cost than someone else. In other words, if you sacrifice less of one good or service to produce another good or service, then you have a comparative advantage. In the example given earlier, Art and Paul are song-writers, but what if both are also capable of performing complex brain surgery? If Art and Paul can both successfully complete two brain surgeries in an hour, then which has a comparative advantage in songwriting, and which has a comparative advantage in brain surgery?

To calculate the comparative advantage, you must determine the opportunity cost that each person faces when producing. In Art's case, for every hit song he writes, he sacrifices two successful brain surgeries. In an hour, Paul can produce either two hit songs or two brain surgeries. This means that Paul sacrifices one brain surgery for every hit song he writes and, therefore, has the comparative advantage in songwriting. Art, on the other hand, has the comparative advantage in brain surgery because for every brain surgery he performs he only sacrifices half of a hit song, compared to Paul who sacrifices a whole hit song for the same surgery. In conclusion, Art should specialize in brain surgery and Paul in songwriting because that is where they find their comparative advantage.

ALERT

It is easy to overlook comparative advantage when determining who should produce what. Just because one person might be more efficient than another does not always mean that that person should be the one doing the task. Remember to always count the opportunity cost.

The theory of comparative advantage allows you to better understand how the American economy has changed over the last sixty years. In that time period the United States transitioned from a low-skilled, manufacturing economy to a high-skilled, diversified economy. Sixty years ago, most of your clothes would have been produced domestically, but today the tags on your clothing indicate that they were manufactured in places as diverse as Vietnam, Bangladesh, Honduras, Morocco, and, of course, China.

In the same time period, a great wave of technological innovation and other cultural advances have taken place. For example, if you were to poll a group of high school freshman about their post–high school plans sixty years ago and another group today, then you would likely discover that today's students are far more likely to pursue higher education than they were in the 1940s and 1950s. In the past, dropping out of high school and working at the mill or the factory was the norm; today, dropping out of high school is cause for concern. There are more jobs as well as more job titles than there were sixty years ago. In other words, there are greater opportunities today than there were sixty years ago. Of course, this comes with one major catch. You must have the education or training in order to take advantage of the opportunity.

So what does this have to do with comparative advantage? An example might help. Consider 100 typical American high school students and then consider 100 young people of the same age in Bangladesh. In which country is the opportunity cost of producing a T-shirt higher? If you look at the American students, then you would have to agree that they have more opportunities than the Bangladeshis. When Americans specialize in T-shirts, more potential doctors, nurses, teachers, engineers, mechanics, fire fighters, police officers, business managers, machinists, and social workers are sacrificed than in Bangladesh, where the majority of workers will most likely become subsistence farmers. The opportunity cost of producing T-shirts is much lower in Bangladesh than in America; therefore, Bangladesh has a comparative advantage in producing T-shirts. Even though the United States has the capacity to produce T-shirts more efficiently (absolute advantage), from an economic standpoint, it makes sense to trade pharmaceuticals, refined chemicals, capital equipment, and know-how for T-shirts.

The Gains from Trade

When trade is both voluntary and free, the buyer and the seller benefit. As an informal proof, consider what happens when you visit the grocery store to purchase milk. As you check out of the store, the cashier often thanks you for your business, and you do something that in other situations is quite peculiar, you thank the cashier right back. Why do you do that, instead of the customary "thank you...you're welcome," exchange? It is because trade benefited both of you. The cashier gets paid, and you get milk without having to milk a cow. This thank you-thank you exchange is typical of voluntary free trade and reminds you that it is a win-win, mutually beneficial exchange.

Because voluntary free trade is mutually beneficial, it is wealth-creating. Be sure to remember this important insight. **Wealth** is nothing more than the collective value of all you own. In an interesting experiment from the Foundation for Teaching Economics, a group of participants were each given a brown paper bag containing a random object. Participants were instructed to keep the object hidden and secretly assign a value to it on a scale from one to ten. Participants recorded the value and submitted the information to the researcher, then the group was given a brief opportunity to reveal their objects and trade freely amongst each other. Soon after the trading occurred, participants were once again asked to assign a value to the object in their possession. As you can probably already guess, the sum of the second set of values is greater than the first. Without anything new being added from the production process, wealth is created through the simple act of voluntary free trade. The experiment not only explains the benefits of free trade amongst individuals in a group, but also the benefits for an entire nation of individuals that trade with individuals in other nations.

FACT

The benefits of free trade were well known to the framers of the United States Constitution. By eliminating internal barriers to trade and making the regulation of interstate commerce a federal function, the United States became one of the largest free trade blocs in history.

International Trade

When you trade with people in other countries, trade becomes **international trade**. The same rationales for domestic trade apply to international trade. Mutual benefit. Win-win. Thank you-thank you. Wealth creation. Yet, international trade is questioned by many. The benefits are often obscured by news reports of job losses due to foreign competition. The pain felt by displaced workers is real and worthy of your attention. However, the benefits of international trade are real and should be considered.

The Case for Free International Trade

Prior to World War II, trade agreements between nations were for the most part bilateral, that is, between the two parties. Under this framework, nations played favorites, which often resulted in unfair treatment for some. The result of all of these bilateral agreements was an international patchwork of high tariffs aimed at protecting certain industries, raising revenue for governments, and meeting the desires of special interests. In the end, the benefits of free trade were not realized and nations drifted toward isolationism and protectionism.

Towards the end of World War II, representatives from much of the industrialized free world gathered in Bretton Woods, New Hampshire, to address the economic issues that were often the cause of international conflict. Although the conference was able to produce the International Monetary Fund (IMF) and the World Bank, it was unable to produce a trade organization for encouraging international cooperation that was satisfactory to the United States. In 1947, many nations, including the United States, came together and formed the General Agreement on Tariffs and Trade (GATT). The goal of GATT was to reduce tariffs and other trade barriers so that member countries could equally enjoy the benefits of free trade. Through a series of negotiations, or rounds, often lasting years, tariffs were significantly reduced and international trade expanded in the latter half of the twentieth century.

ESSENTIAL

Several organizations have been developed to encourage free international trade. The European Union (EU) and the North American Free Trade Agreement (NAFTA) are notable examples. Both have been effective in increasing trade among their various members.

The growth in international trade was accompanied by a rise in living standards amongst the members of the agreement. In 1995, with a successful end to the Uruguay Round of GATT negotiations, the GATT became the World Trade Organization (WTO). Under GATT and later the WTO, more and more countries have become supporters of lower tariffs and fewer barriers to trade. As a result, international trade has continued to expand and many nations have reaped the benefits. For example, since joining the EU and opening itself to international trade, Ireland has gone from being one of Europe's poorest countries to one of its wealthiest.

The Case Against International Trade

Free international trade has many detractors. Among these are environmentalists, labor unions, human rights activists, and politicians concerned with sovereignty. Many sincere individuals concerned with the environment have reservations about international trade. The chief concern is that as countries specialize, production will concentrate in countries that have fewer regulations to protect the environment from pollution and habitat destruction. Labor unions oppose free trade on the grounds that production will shift towards low-wage countries that have little or no union representation, and therefore negatively impact their membership. Human rights activists often oppose free trade as production shifts toward countries where working conditions are miserable and often inhumane, and where workers are not afforded the same rights and privileges as in industrialized nations. Politicians and their constituents concerned with loss of national sovereignty often oppose free trade agreements on the grounds that decisions affecting the nation are being made by an international body not directly subject to the people. All of these detractors voice their concerns and have become influential in WTO negotiations.

Other Bases of Trade

In addition to comparative advantage, other theories have been proposed to explain how and why people in different countries trade. These theories have expanded on David Ricardo's theory of comparative advantage and help to better explain the trade patterns that exist today. Eli Heckscher and Bertil Ohlin proposed a theory in the 1920s that looks at the differences between countries with respect to their available capital and labor. Paul Krugman proposed a theory that better reflects the realities of trade today and is based on a country's preference for diversity in consumption and industry's tendency to cluster.

The **Heckscher-Ohlin theory** says that if two countries have different mixes of labor and capital and specialize according to their mix, then they will benefit more from trade than two countries with similar mixes of labor and capital. As an example, assume that Vietnam has much labor but little capital and that Germany has much capital but not as much labor. Further assume that clothing manufacturing is labor-intensive and that automobile production is capital-intensive. Based on Heckscher-Ohlin theory, if Viet-nam specializes in clothing and Germany in cars, and both countries trade, then the gains will be greater than if they had traded with countries that were more similar to their own.

Paul Krugman looked at trade patterns between countries and observed that Ricardo's theory of comparative advantage and the Heckscher-Ohlin theory did not always explain existing international trade patterns. For example, both Italy and Germany produce cars, and yet they freely import and export cars with each other. This allows consumers to enjoy some diversity. Sure, you might be a proud German, but there is just *something* about a Ferrari.

FACT

Since the late 1960s, worthy economists have been awarded the Sveriges Riksbank Prize in Economic Sciences in Memory of Alfred Nobel. Both Bertil Ohlin and Paul Krugman were awarded the prize for their work on international trade theory, with Ohlin receiving his in 1977 and Krugman in 2008.

Krugman also observed that much of the trade that occurs between nations stems from intraindustry trade. Industries tend to cluster together. For example, computer motherboard manufacturers in Taiwan may trade with computer manufacturers in Malaysia and China, which in turn export complete systems to Japan and the United States.

A Look at Trade Barriers

From time to time, countries will seek to tax, limit, or even ban international trade. Why would governments do this if voluntary trade is mutually beneficial? The answer lies in the fact that even though voluntary trade is mutually beneficial, the benefits are spread out over society, but the costs are sometimes borne directly by a specific group. People might have a strong interest in preserving their industry, raising tax revenue, saving the environment, or even creating social change. At times a country might limit trade in order to punish another country. Tariffs, quotas, and embargoes are a few of the tools that a country will use in order to accomplish these other interests.

Tariffs

A **tariff** is a tax on trade. Tariffs can be used to raise revenue for the government or in order to benefit a certain segment of the economy. You might pressure Congress to enact a tariff on imports if your industry is subject to foreign competition. For example, for years the United States' steel industry was protected from cheap foreign competition by protective tariffs. In 2007, India proposed a tariff on rice exports in order to prevent food shortages. The Smoot-Hawley Tariff Act of 1930 was intended to protect American industry and raise much needed tax revenue for the government.

ESSENTIAL

The United States Constitution has some things to say about tariffs. In Section 1 Article 8, Congress is given the power to collect a tariff on imports, but in Section 1 Article 9, Congress is forbidden from placing a tariff on exports.

Tariffs are not without their downsides. Protective tariffs often have the effect of preventing competition and encouraging waste and inefficiency. Revenue tariffs often fail to raise tax revenue because people might stop buying the now expensive imports. Export tariffs might give producers an incentive not to produce. The Smoot-Hawley tariff had the effect of not only reducing foreign imports but also of reducing exports as other countries established tariffs in retaliation, preventing the much-needed tax revenue from materializing. Before politicians consider establishing a tariff, it would be wise to look back at history and see if there might be some unintended consequences.

Quotas

Quotas are limits on trade. Instead of a tax on imports, you might use a quota to limit the number of imported goods coming into your country. In the 1970s and 1980s, U.S. automobile manufacturers and labor unions supported government quotas on foreign car imports to limit competition and preserve American jobs. The result was higher prices and lower quality. Eventually, Japanese and German firms bypassed the quotas by establishing their factories in the United States. In the end, domestic producers faced more competition at home and labor unions suffered as foreign firms established their factories in states where unions had less power.

Quotas create other problems as well. They do not generate tax revenue for the government, but do create more responsibility. They provide an incentive to smuggle goods illegally in order to avoid the quota, thus creating black markets. In addition, quotas may be manipulated by foreign firms to limit competition from other foreign firms. For example, if there is a quota on German cars imported into the United States, then the German firm that first fills the quota has effectively blocked other German firms from competing in the American market.

Embargoes

An **embargo** is a ban on trade with another country. The purpose of an embargo is usually to punish a country for some offense. The embargo you may be most familiar with is America's embargo against Cuba. In the wake of the communist revolution, and later the Cuban Missile Crisis, the United States enacted an embargo that banned all trade with the island nation. Even though the events are now far in the past, the embargo persists. Once again, you might consider who benefits from the trade embargo in order to understand why it is still in place.

CHAPTER 3: That's Not the Way We Do Things!

Why are some countries so rich and others so poor? Does the presence of abundant natural resources account for a country's wealth? Why is there such a lack of economic development among different indigenous groups around the world? How important is government to an economy, and what are the government's appropriate economic roles? If you are interested in understanding what these answers might be, then a study of the different economic systems will shed some light.

Different Economic Systems

Regardless of the time in history or the geographic location, every society has had to and will continue to deal with the fundamental problem of scarcity. In order to survive, society must make decisions about how to best use its scarce resources (land, labor, capital, and entrepreneurial ability). Economists have concluded that for societies to survive with their limited resources, they must answer three basic questions: What to produce? How to produce? Who to produce for?

FACT

> The eighteenth century, also known as the Age of Reason, or the Enlightenment, saw a fundamental shift in the way

people viewed their world. The year 1776 was especially important, for it not only was the year that Thomas Jefferson wrote the Declaration of Independence, but it was also the year that Adam Smith's *An Inquiry into the Nature and Causes of the Wealth of Nations* was published.

Throughout history people have developed a variety of systems to answer these questions. Most primitive societies developed what economists refer to as traditional economies. With the development of civilization came command economies, and following the Enlightenment, market economies finally emerged. In addition, combinations of these primary systems developed such as communism, socialism, and capitalism.

Traditional Economies

What do the Inuit, the Bushmen of the Kalahari, Aboriginal Australians, Costa Rica's Bri Bri, and the Dani of New Guinea have in common? If you guessed climate, then you would be way off. Instead, this list of indigenous groups from around the world is just a sample of the people who have practiced or continue to practice **traditional economy**.

In a traditional economic system, the questions of what and how to produce and who to produce for are answered by tradition. If ever you have witnessed a documentary on a primitive culture, then you have also seen a traditional economy in action. The Kalahari's Bushmen live in one of the world's harshest environments where even the most basic resources are in meager supply. In order to survive and have enough food, the Bushmen have developed a division of labor based on gender. Women perform the food gathering and men perform the hunting. The food is then shared with the whole tribe. In this type of system, stability and continuity are favored over innovation and change. The roles of the people are defined by gender and status in the community. In this system, the old, young, weak, and disabled are cared for by the group. The group shares the few possessions they have, and private property is an alien concept. For the most part, everyone in this system understands his or her relationship to the community, and as a result, life hums along in a fairly predictable way.

ALERT

As the world's indigenous people are absorbed into modern economies, they often suffer. The skill sets and practices that help them to survive in the rainforest, desert, or tundra are rendered useless in a world of freeways, factories, and fast food.

Command Economies

As hunter-gatherer societies grew and eventually exhausted their natural food supplies, some survived by becoming sedentary farmers. With the advent of farming came a need for an organized system of planting, harvesting, and storing crops. This required a greater amount of structure than existed in a traditional economy. In order to ensure the survival of the society, decisions had to be made about what crops to grow and how much of the harvest to store. Over time, decision-making became centralized and the **command economic system** developed. The key characteristic of the command economy is centralized decision-making. Either one or a group of powerful individuals makes the key economic decisions for the entire society.

Examples of command systems include most, if not all, ancient civilizations, plus the communist countries of today. The pharaohs of Egypt represent the centralized decision-making present in a command economy. The pharaoh and his various officials made the key economic decisions of what to produce, how to produce, and for whom to produce. The decisions might have gone something like this, "I command you to construct a big pyramid of brick and mortar using slaves for labor, and all of it is for me." The advantage of this type of system is the ability for decision-makers to produce rapid changes in their society. For example, Soviet dictator Josef Stalin's five-year plans quickly transformed the Soviet Union from a peasant-based agrarian society into one of the world's industrial superpowers.

FACT

During World War II, the United States practiced command economy when the government took over factories and planned production for the war effort. Every aspect of American life was in some way influenced by government involvement in the economy. Even today you can see the influence. The modern payroll withholding system was instituted during the war to provide the government with a steady stream of tax revenue.

History reveals the tragic downside of command economic systems. As mentioned above, the pharaohs used slave labor, and Stalin's five-year plans were only accomplished through the forced relocation of millions of people and at the cost of millions of lives. Rarely do the decision-makers meet the wants and needs of the common citizen.

The state, acting as a sole producer without competition, has little incentive to provide anything more than basic necessities. As a result, no variety exists in a command economy. In addition, the citizens serve for the purpose of the economy and state, as opposed to the economy and state serving the citizens. North Korea is a perfect example. Property belongs only to the state. Many workers have little personal incentive to produce, and those that do may have little regard for quality. Individuality, innovation, and variety are completely lacking in the command system.

Hollywood illustrates this lack of choice in the 1984 movie *Moscow on the Hudson*. A Soviet saxophone-playing defector, portrayed by Robin Williams, is shown shopping in a Moscow state store, where there is a lack of food, long lines, and no choices. Later, after defecting from the visiting Moscow Circus in a Manhattan Bloomingdale's, he goes to an American supermarket. The character proceeds to pass out from the overwhelming number of choices he faces when his adopted family sends him to buy coffee.

Market Economies

In total contrast to the command economic system is the **market economy**. Market economies are characterized by a complete lack of centralized decision-making. As opposed to top-down planning, market economies operate bottom-up. Individuals trying to satisfy their own self-interest answer the questions of what, how, and for whom to produce. Instead of tradition or government mandate, the "invisible hand" of the market directs resources to their best use. Private citizens, acting on their own free will as buyers or sellers, trade their resources or finished products in the market in order to increase their own well-being. Though it might appear counterintuitive, market economies achieve greater abundance, variety, and satisfaction than either traditional or command economic systems.

QUESTION

Does an abundance of natural resources make a country wealthy?

Not necessarily. Resource-poor Japan is five times wealthier than Russia, which is the world's largest producer of oil and diamonds. Economists refer to this type of discrepancy as the paradox of oil. Countries that often have abundant natural resources invest heavily in extracting its resources at the expense of investing in their people.

Although they cannot be classified as pure market systems, Hong Kong, the United States, Australia, and New Zealand are representative of market economies. In each you will see a greater variety of goods and services being produced than anywhere else. Also, because the focus is not on serving the state, individuals are free to choose their vocation, own private property, and determine for themselves how to best use the resources they possess. Markets reward innovation, productivity, and efficiency but discourage complacency, idleness, and waste. If markets have a downside, it is that those who are unable or unwilling to produce because of either circumstance or choice are often sidelined and unable to enjoy the benefits of the system.

Capitalism Versus Socialism

Today, traditional economies are few and far between, command economies are waning, and pure market economies are nonexistent. What does exist throughout much of the world are varying combinations of command and market systems; in effect, economic hybrids. The two most common economic hybrids are **socialism** and **capitalism**. Imagine an economic continuum with a pure command economic system on the left and a pure market system on the right. If you were to arrange modern nations along this continuum, towards the far left would be places like North Korea and Iran, in the middle would appear many western European and Latin American nations, and to the right would appear many former British colonies, such as the United States, Australia, and Hong Kong. For all practical purposes, those nations on the left were described in the discussion on command economies. However, the middle and the right of the continuum represent the dichotomy of socialism and capitalism.

ALERT

Do not confuse capitalism with democracy. The two do not necessarily go together. India is the world's largest democracy, but it is considered a socialist economy. Hong Kong has never really experienced democracy and yet it is the epitome of capitalism.

The difference between these two systems lies in the degree of government influence and state ownership of the factors of production. Countries that are capitalist rely on market prices for efficient product allocation, promote the private ownership of economic resources, and leave most economic decisions to individuals. They do, however, permit for government to regulate markets, preserve competition, subsidize and tax firms, enforce private contracts, and redistribute income from workers to nonworkers. For example, the United States government creates rules for the labor market, breaks up monopolies, subsidizes corn growers, taxes polluters, hears cases involving breaches of contract, and collects social security taxes.

In socialism, the government takes a much more active role in the economy. Although individuals are allowed private property, the state may own firms in key industries and regulate even more economic decisions than in capitalism. In France it is not uncommon for the government to take a major stake in French companies, if not outright own them. The French labor market is more heavily regulated than its American counterpart. In 2006, French students poured into the streets, protesting the fact that the government was being pressured by French firms for the right to fire employees at will during their first two years of employment. Compare this to the United States where there is no guarantee of employment.

FACT

British Prime Minister Margaret Thatcher is credited with reversing the United Kingdom's drift towards socialism. With the end of World War II, the British had moved towards socialism with the nationalization of several key industries. While in office, she began the process of privatization where state-owned companies were sold to private shareholders.

More often than not, socialist countries manage the prices of many goods and services. The EU manages prices on such things as pharmaceuticals, cell telephone service, and food. Also, socialist countries are more active in taxing in order to redistribute income from workers to nonworkers. Germany is well known for its generous cradle-to-grave social welfare system that promises care for its citizens. The German welfare state is financed by a redistributive tax system that many Americans would find intolerable. As of 2008, the highest marginal tax rate on personal income in Germany was 45% compared to the United States rate of 33%.

Adam Smith, Karl Marx, and Human Nature

According to John Maynard Keynes, "The ideas of economists and political philosophers, both when they are right and when they are wrong, are more powerful than is commonly understood. Indeed, the world is ruled by little else. Practical men, who believe themselves to be quite exempt from any intellectual influence, are usually the slaves of some defunct economist." Keynes's insight into the influence of economic thought on the lives of people can be seen in the various economic systems that have developed over time. In one camp are those who would have the state as the primary caretaker of the people. In the other are those who believe that the problem of scarcity can only be addressed through individual economic freedom.

Karl Marx said, "From each according to his abilities, to each according to his needs." Marx envisioned an economy where the problem of scarcity was addressed through the complete redistribution of wealth and income, from the owners of land and capital to the workers. In his utopian vision, social justice, economic equality, and relief from scarcity would be achieved when society was organized in such a way that all were equal regardless of their level of productivity.

According to Adam Smith, "It is not from the benevolence of the butcher, the brewer, or the baker, that we expect our dinner, but from their regard to their own self-interest. We address ourselves, not to their humanity but to their self-love, and never talk to them of our own necessities but of their advantages." Adam Smith had a different vision of society. In his view, productivity was the determinant of wealth, and rational self-interest was the motivating force that would provide society a means of escape from scarcity. Adam Smith believed that when society harnessed the power of self-interest, the greatest good could be accomplished.

These two men had two very different ways of addressing the fundamental problem of scarcity. Which do you think better captures human nature? If you believe that people are basically good and seek to serve one another, then the ideas of Marx ring true. However, if you believe that people are at heart selfish and pursue their own ambition, then the words of Adam Smith might appear more valid. Regardless of what you believe, both Smith and Marx have managed to influence society in ways that are still evident today.

CHAPTER 4: The Story of Money

Of all humankind's inventions, money stands out as one of the most widespread and useful. A day probably does not go by that you don't use it or think about it. It's hard to imagine a time when people did not have money, and it can be scary to imagine what your life would be without it. From barter, to shells, to coin, to paper, to digital, the story of money spans much of human history.

Barter

Before money was invented, and in times when money was either worthless or extremely scarce, **barter** was used as a means for people to get what they needed or wanted. Barter is simply the act of exchanging one good or service for another good or service. An example of barter is when a farmer trades a dozen chicken eggs with the baker for a fresh loaf of bread. Although barter was more common in the past, it still exists today. In order to evade taxes, some businesses will barter their services with other businesses. To see why this happens, consider the following example. Assume that a dentist needs her car fixed and that an auto mechanic is in desperate need of a root canal. Instead of charging each other for their services and creating taxable income, both can benefit from barter. You should know that the Internal Revenue Service considers barter every bit as taxable as cash income, but that does not seem to stop many from continuing the practice.

Barter is not without its downsides. Obviously, trade will not occur unless both parties want what the other party has to offer. This is referred to as the double coincidence of wants. In the example of the farmer and baker, if the baker has no need or desire for eggs, then the farmer is out of luck and does not get any bread. However, if the farmer is enterprising and utilizes his network of village friends, he might discover that the baker is in need of some new cast-iron trivets for cooling his bread, and it just so happens that the blacksmith needs a new lamb's wool sweater. Upon further investigation, the farmer discovers that the weaver has been craving an omelet for the past week. The farmer will then trade the eggs for the sweater, the sweater for the trivets, and the trivets for his fresh baked loaf of bread. Whew! There has got to be an easier way to do things.

The Development of Money

The previous example illustrates the need for a more efficient means of exchanging goods and services. As a result of the downsides to barter, cultures in different times and places eventually developed money. Consider cattle. Cattle were immensely useful in the past. They provided meat, milk, leather, and if harnessed, power. Throw in the fact that their dung could be used for fuel or fertilizer, and it becomes obvious that to the people of the past, cattle were as good as gold. Cattle served as one of the earliest forms of money because it seemed everyone could use them. Because cattle were universally accepted, they served as an efficient means of exchange. Now, before you consider trading in your dollars for cows, consider that your dollars do not wander off, nor must you feed them. Add to that the fact that cows do not fit in the average-sized purse, and you can see that money needed to evolve further.

Functions and Characteristics of Money

Regardless of the form it takes, **money** is anything that functions as a medium of exchange, store of value, or standard of value. **Medium of exchange** means that money is being used for the purpose of buying and selling goods or services. Money acts as a store of value when you get it today and are still able to use it later. Money is functioning as a standard of value when you are using it to measure how much a good or service is worth. When you write a check for a friend's wedding gift, then you are using money as a medium of exchange. Imagine that you go to the wedding in June and slip fifty dollars into your suit pocket for some spending cash. Now imagine that it is November of that same year and you are now at your friend's divorce hearing (things did not work out). In the midst of the hearing you reach your hand into your pocket and shout for joy, "Store of Value!" Although you feel awkward, you are pleased because the fifty dollars is still worth fifty dollars. After the scene you caused in the courtroom, you discover that court-ordered sensitivity training is going to cost you $150 per session. That would be using money as a standard of value.

Money works best when it is has the following characteristics: portability, durability, divisibility, stability, and acceptability. **Portability** refers to the ease with which money can be carried from place to place. It is desirable for money to be **durable** so that when you forget to remove it from your pocket before doing the laundry, you do not wind up broke. All of that annoying change that seems to find its way between the cushions of your couch is proof that money is **divisible** into smaller, peskier units. **Stability** exists when money's value does not vary too much. **Acceptability** helps to explain why you are far more likely to stop and pick up a ten dollar bill as opposed to ten dollars in Monopoly money.

Money Throughout History

Across time and cultures, many things have served as money, including salt, tobacco, shells, large stones, precious and nonprecious metals, leather, and cigarettes, to name a few. No matter the form, the purpose of money is to serve as a medium of exchange, store of value, and act as a standard of value. Over time, money has evolved from commodity money to representative money, and now to inconvertible fiat.

FACT

The Pacific island of Yap is known for its money, which is decidedly not portable. Large rounded stones weighing hundreds of pounds are used as a medium of exchange. If you plan on visiting Yap, wait until you are there to exchange your currency because it will not fit under the seat or in the overhead compartment of an airliner.

When relatively scarce minerals, metals, or agricultural products are used as a means of exchange, they are considered **commodity money**. Gold and silver struck into coins are examples of commodity money. An advantage of commodity money is that it can be used for purposes other than money. In the 1980s, many women adorned themselves in jewelry featuring gold coins such as the Chinese Panda or the Canadian Maple Leaf. American colonists not only smoked tobacco, they also used it as money. The salt we take for granted was at one time scarce enough that Roman soldiers were paid in it. On the other hand, a commodity's usefulness also a makes it a disadvantage to using it as money. If a country is dependent upon using a commodity for its money and as a resource, then money may be too precious.

Representative money developed as an alternative to commodity money. One of the properties of gold is its high density. Transactions requiring large amounts of gold would have been extremely heavy and difficult to transport. Goldsmiths offered a solution to this problem. By issuing receipts for gold they had on deposit, representative paper money was born. Instead of trading the physical gold, all people had to do was trade the receipts for the gold. Whenever they wanted the actual gold, they could redeem their receipts. After years of acceptance, people became more comfortable with the idea of representative paper money and the concept stuck.

Because people were already familiar with representative paper money, then the next step in the evolution of money is not all that difficult to understand. Why bother making the paper money redeemable in anything? Several times in history, the convertibility of representative money into gold or silver had been halted because of war or other crises. In 1933, President Franklin Delano Roosevelt signed an executive order that transformed the dollar from being a form of representative money in to what is called **inconvertible fiat**. Inconvertible fiat refers to both paper and virtual money that is intrinsically worthless and is not redeemable or backed by some real commodity. It is money because the government says so *and* we are willing to accept it. The United States dollar, the euro, the pound, the yen, and most other world currencies meet the definition of inconvertible fiat.

The Gold Standard

Today you might hear people talking about doing away with the Fed, America's central bank, and returning the United States to a **gold standard**. Under a gold standard, money is representative money backed by a fixed amount of gold. For example, throughout much of the early twentieth century, the United States was on a gold standard, which meant that your dollars could be traded in at the bank for either a fixed amount of silver or gold depending on the certificate you held. The gold standard allows a country to only print as much currency as there is gold or silver to back it up. Like any system, there are both advantages and disadvantages to a gold standard.

One clear advantage of a gold standard is that it acts to prevent a government from overprinting and devaluing the currency. This greatly reduces the chance for inflation to creep into the economy and also acts as a check against governments borrowing too heavily. **Inflation** is ultimately caused by too much money in circulation. This excess money drives up prices and makes most things more expensive. Under a gold standard people's expectations of inflation are held in check, and that leads to not only greater price stability but also serves to stabilize employment and the economy as a whole.

The gold standard's major disadvantage is that it acts as a limit on economic growth. According to economist Adam Smith, wealth is not a function of how much gold or silver a country has but is rather the sum of all the goods and services an economy produces. It makes sense that the amount of money an economy has should in some way reflect its capacity to produce wealth. As businesses expand, they require money in order to purchase the tools, factories, and equipment necessary to meet both their productive needs and the demand for their goods and services. As an economy's productive capacity grows, then so should its money supply. Because a gold standard requires that money be backed in the metal, then the scarcity of the metal constrains the ability of the economy to produce more capital and grow.

Welcome to the Matrix

The inconvertible fiat standard that exists today addresses the weakness of the gold standard. Because it is not backed by anything real or tangible, the money supply is able to grow as the economy grows. This flexibility explains why the inconvertible fiat standard persists. So that people do not lose confidence in the money supply, the central bank of a country must carefully control its availability so that it does not become too plentiful or too scarce.

FACT

> Money is debt. The United States dollar is a promise to pay from the United States Federal Reserve to the holder. You might ask, "A promise to pay what?" The answer is another dollar.

When you stop to think about it, the inconvertible fiat standard sounds like science fiction. Today's money is intrinsically worthless and is only redeemable for more of the same. The system works because the government says so and everyone collectively believes it. Money is backed by nothing more than faith. When you think about direct deposit, online bill payments, debit cards, and checks, the idea of money is weirder still. You work, pay your bills, buy your groceries, and manage to survive and even thrive in the economy, yet you can go for days or weeks without even touching, seeing, or smelling money. Money is imaginary. Ponder your bank account. There are not little stacks of dollar bills sitting in the bank vault with your name on them. Instead, checking and savings accounts are nothing more than information stored on computers.

Because money is backed by faith, anything that erodes that faith is destructive to money. Overprinting and counterfeiting pose a serious threat to any money supply. Overprinting is dangerous because it causes the money to lose value and is therefore highly inflationary. During the inter-war period, the German Weimar Republic overprinted their currency, and this led to rampant inflation and financial ruin. Counterfeiting occurs when either criminals or enemy governments copy currency and then distribute it. If successful, counterfeiting acts to disrupt an economy by causing people to question the currency. In response to these dangers, governments are constantly coming up with new ways to better secure their money and make it more difficult to counterfeit.

M1 and M2

This strange stuff called money is managed and measured by the Federal Reserve in the United States. There are two primary measures that the Fed uses when describing the money supply, M1 and M2. The **M1** is composed of all of the checking account balances, cash, coins, and traveler's checks circulating in the economy. The **M2** is composed of everything in the M1 plus all savings account balances, certificates of deposit, money market account balances, and U.S. dollars on deposit in foreign banks. The M1 is mainly used as a medium of exchange, whereas the M2 is used as a store of value. The M2 is larger and less liquid than the M1.

ESSENTIAL

The currency and coin inside of a bank is not counted in the M1. Why not? Because it's not money until you walk out of the bank. So technically, it is inaccurate for a bank robber to demand money while in a bank.

Changes in the M1 and M2 are monitored by the Fed and act as indicators of economic activity. Sudden changes in the ratio of M1 to M2 might indicate either imminent inflation or recession. In general, if the M1 grows faster than the combined rate of labor force and productivity growth, then inflation will result. If, however, the M2 were to suddenly grow at the expense of M1 because people are saving and not spending, then that would tend to indicate that the economy is headed toward recession.

The Time Value of Money and Interest Rates

Economists observe that money's value is affected by time: a dollar today is worth more than a dollar tomorrow. This has to do with opportunity cost and inflation. If you lend your friend the money in your wallet, your opportunity cost is the sacrifice of its immediate use. When your friend eventually pays you back, the money will have lost purchasing power due to inflation. As a result, when people lend money, they often ask to be rewarded with interest to offset the opportunity cost and inflation. When you deposit money in your savings account, you expect interest for the same reason.

Interest is nothing more than a payment for using money. An **interest rate** is the price of using money. What determines this price? It helps to think of an interest rate as set of blocks stacked upon each other. The first block represents the opportunity cost of using money. Some people will readily forgo the immediate use of their funds in order to receive interest. Others might be unwilling to sacrifice the immediate use of their money. In the absence of inflation or risk, the interest rate that equates the level of saving to the level of borrowing is the basic interest rate or **real interest rate**.

ALERT

Many older Americans can remember a time in the late 1970s and early 1980s when interest rates were as high as 20%. Compared to today's low rates, that is quite a difference. The explanation for the difference is inflation. Interest rates include a premium for inflation. So even though your savings is earning a much lower rate today, when you adjust for inflation, the differences pretty much disappear.

The second block of interest represents the cost of **expected inflation**. Assume that inflation has been stable for years at a rate of 3% and people are pretty confident that it will remain at 3%. A lender or investor will cover the cost of expected inflation and add 3% to the 2% real interest rate, to arrive at a nominal interest rate of 5%. The nominal interest rate is the basic rate that an investor or lender will charge for the use of money.

If there is a chance that the loan or investment will go bad, then it makes sense to add another block to the stack. This third block is referred to as **default risk premium**. The bigger the risk of default or nonpayment, the bigger the block, and the bigger the total nominal interest rate.

If there is a chance that the investment or loan will be difficult to turn around and sell to another lender or investor, like a ten-year car loan, another interest block is added. Not many people are willing to assume the risk of buying a loan that is backed by a fully depreciated asset. This fourth block is referred to as a **liquidity premium**.

One final block is added for **maturity risk**. As time passes, there is a chance that interest rates will increase. If this happens, the value of the investment decreases because who would want an investment that earns only 2% when you can get a similar one that earns 4%?

So, here's an example of how an interest rate is calculated: the real interest rate is 2% and the inflation rate is 3%, which combine to form a 5% nominal interest rate. Assume the risk premium is 4%, the liquidity premium is 2%, and the maturity risk premium is 1%. The total nominal interest rate would stack up to 12% (2% + 3% + 4% + 2% + 1%).

CHAPTER 5: Banks

Hushed tones, cool marble counters, the smell of cash, pens chained to tables with blank deposit slips, solid steel doors with impressive locks, velvet-lined cords directing customers to the appropriate teller — a bank is an important place. Banks are everywhere. From small towns to large cities, the ubiquity of banks reveals their importance to the economy. Much maligned of late, banks are an integral part of the economy. Without them, capitalism would not function.

The Origin of Banking

The roots of banking can be traced to the earliest civilizations. The Egyptians and early societies of the Middle East developed the prototype upon which modern banking is based. Agricultural commodities were stored in granaries operated by the government, and records of deposits and withdrawals were maintained. Ancient civilization introduced the money changers, who would exchange currency from different countries so that merchants, travelers, and pilgrims could pay taxes or make religious offerings.

FACT

> The name "bank" comes from the Italian word banco. A banco was a table covered in green cloth where

moneylenders completed their transactions in the marketplace. Those are pretty humble beginnings for a major industry.

In the Renaissance era, Italian city-states were home to the first banks, which financed trade, the state, and the church. In order to avoid the church's prohibition against **usury** (charging interest), the bankers would lend in one currency but demand repayment in another currency. Profit was thus earned by using different exchange rates at the time the loan was made and when it was repaid. The successes of the Italian bankers induced a spread of banking further across the continent. In England, goldsmiths were responsible not only for storing gold and issuing receipts but also for developing what is now termed fractional reserve banking. By issuing more receipts than there was gold on deposit, the goldsmiths increased the profit potential of the banking industry.

From the time of the American Revolution to the Civil War, the United States saw an expansion of relatively unregulated banking that helped finance the growth of the young republic. Modern banking in the United States traces its origins to the National Bank Act of 1863, which gave the government a means to finance the Civil War.

The Function of Banks

Banks serve a variety of functions in the economy. They act as safe places for people to store their wealth, they help to facilitate trade by providing alternative methods of payment, but most importantly, they bring together savers and borrowers. Each of these functions is critical to the smooth operation of the economy.

In John Locke's *Second Treatise of Civil Government,* he observes that in order for people to enjoy their private property they must be secure in that property. Thomas Jefferson paraphrases Locke in the Declaration of Independence when he states: "we are endowed by our Creator with certain unalienable rights, that among these are life, liberty and the pursuit of happiness." What Jefferson refers to as happiness, Locke refers to as property. Although Locke and Jefferson were not referring to banks but governments, banks do serve an important function in a free society by providing a safe place for people to store their wealth. When you know that your private property is secure, then you are better able to enjoy your freedoms.

ESSENTIAL

Historians trace the development of the check to the Knights Templar and Hospitallers who used signed documents in order to transfer wealth among their order's houses.

By providing their customers with check-writing privileges, debit cards, credit cards, and access to cashier's checks, banks help to facilitate trade. With multiple means of access to stored wealth, consumers are able to make purchases more often and in more places. This allows businesses to employ more land, labor, and capital, which in turn results in a fully employed economy.

Acting as a go-between or intermediary between savers and borrowers is probably the most important function of banks. Banks induce people to save their money by offering to pay interest. These savings are then lent to borrowers at an interest rate higher than that paid to savers, allowing the bank to profit. The saver benefits because he earns interest on a safe and relatively liquid financial investment without having to evaluate whether a potential borrower is a good risk. The borrower benefits by having access to a large pool of funds. This is important to the economy because borrowers can now purchase durable goods or invest in capital or housing, which creates jobs and leads to economic growth.

Balance Sheets

To understand the economic importance of banks, it helps to look at a bank's balance sheet. A **balance sheet** compares the assets a bank owns with the liabilities it owes. If you have never taken a course in accounting, then you might not be familiar with the following equation: Assets = Liabilities + Stockholder's Equity. A bank's **assets** include buildings, equipment, loans to customers, treasury securities, vault cash, and reserves. The **liabilities** of a bank include customer's deposits and loans from either other banks or the Fed. **Stockholder equity,** or a bank's financial capital, is the ownership interest in the bank represented by shares of stock. Because assets equal liabilities plus stockholder's equity, changes in a bank's liabilities can create an equal change in a bank's assets. For example, if customers deposit $100,000 in a bank (a liability), then the bank's reserves increase by $100,000 (an asset). On the other hand, if customers withdraw $25,000, then bank reserves are reduced by $25,000 as well.

QUESTION

What is a capital requirement?

Banks are required by law to maintain capital requirements. The purpose of the capital requirement is to ensure that the bank is able pay depositors if some of the bank's borrowers are unable to repay their loans. Capital requirements also mean that banks have a vested interest in making sound loans.

What exactly are bank reserves? **Reserves** are funds that are either available for lending or held against checkable deposits. The reserves available for lending are called excess reserves, and those held against checkable (but not savings) deposits are required reserves. Required reserves are held either as cash in the bank's vault or are deposited in the bank's reserve account with the Fed. In the United States, the **required reserve ratio**, set by the Fed, is the percentage of checkable deposits that a bank cannot lend. For large banks, the required reserve ratio is 10%. Therefore, assuming a 10% required reserve ratio, if customers deposit $100,000 into checking accounts, required reserves increase by $10,000 and excess reserves increase by $90,000. When the economy is healthy, banks tend to lend out all excess reserves. Why? Banks profit by charging interest on loans, so they have a strong incentive to maximize the amount they lend.

How Banks Create Money

Contrary to popular belief, most money is not created on government printing presses. When banks accept deposits and make loans, money is created. For example, a $100,000 checking deposit generates an increase in excess reserves of $90,000. If the bank lends the full $90,000 to a customer who in turn purchases a recreational vehicle, the seller of the vehicle might then deposit the

$90,000 in the bank. What happened to the checkable deposit balance in the bank? It grew from $100,000 to $190,000 in a short period of time. Money was created. The process does not stop with just this transaction. You can see that the bank now has $90,000 in new deposits. The bank will hold 10% as required reserve and lend the rest. The proceeds of the loan will be redeposited, and now $81,000 of new money is created. This process continues until all excess reserves are loaned out.

FACT

The presence of excess reserves in banks does not necessarily indicate that banks are refraining from making loans. In response to the credit freeze of 2008, the Fed instituted several emergency measures that had the effect of greatly increasing excess reserves in the banking system.

Economists are able to estimate the growth in the money supply with the **money multiplier**, which is one divided by the required reserve ratio. Given that the reserve ratio is 10%, the money multiplier is ten. If $30,000 is deposited into a checking account, economists would predict that the money supply will grow by a maximum of $270,000 ([$30,000 − $3,000] x 10). The accuracy of the money multiplier as a predictor of the money supply is constrained by two factors. The multiplier assumes that banks lend all excess reserves and that the loans are all redeposited. If either of these assumptions does not hold, the multiplier effect is reduced. Many Americans hold on to their cash and that acts as a limit to the money multiplier. Just as easily as money is created, money can also be destroyed. Remember, money is created when customers make deposits and banks make loans. Money is destroyed when customers withdraw balances and pay off loans. Consider the following example. If Maria writes a $10,000 check to pay off her car loan, checkable deposits are reduced by $10,000, and the money supply shrinks.

Banks as a System

Banks work together as a system in bringing together savers and borrowers. They accomplish this by lending and borrowing directly from each other. If a bank is low on reserves and will not fulfill their daily reserve requirement, they are able to borrow from other banks overnight in the **fed funds market**. For example, assume that Bank X has a customer who withdraws her savings at the end of the business day. Because banks do not hold reserves against savings deposits, this might leave Bank X without the required reserves it must hold against checking deposits. Bank Y, however, may have excess reserves available only earning minimal interest in their reserve account with the Fed. For Bank Y, it is profitable to lend its excess reserves to Bank X at the higher fed funds rate. The **fed funds rate** is the interest rate targeted by the Fed that banks charge each other for the overnight use of excess reserves.

QUESTION

What's LIBOR?

LIBOR stands for the London InterBank Offered Rate and is the market interest rate for overnight interbank loans.

Unlike the fed funds rate, which is manipulated by the Fed, LIBOR is independent of government influence.

Sometimes banks may have excess reserves, but businesses or households may not be willing to borrow. Assume that Bank East is holding excess reserves but has no opportunities to lend in its region. Bank West has no excess reserves but has businesses and consumers clamoring for loans. Bank West can borrow from Bank East in the fed funds market and provide loans for its customers. Bank East profits by earning the fed funds rate, and Bank West profits by earning the higher interest rate it charges its customers.

Bank Runs

Bank runs or bank panics have occurred multiple times throughout American history. Because banks operate with far less than 100% required reserves, it is possible that if enough customers demand their account balances on a single day, the bank will not be able to meet the demand. This of course would be catastrophic for the bank and its customers. The bank would be insolvent, and the customers unable to withdraw their funds would be broke.

In the classic film *It's A Wonderful Life*, George Bailey, played by Jimmy Stewart, heads off a bank run at his family's Building and Loan by explaining a bank balance sheet and fractional reserve banking.

What would cause customers to demand their account balances all at once? Fear, whether based in truth or not. Many bank panics have been caused because of rumor or speculation about a bank's financial health. If enough people believe the rumor, they will logically want to withdraw their funds and move them to either their home or another financial institution. Once the line starts forming at the bank's door, other customers will notice and the rumor will spread. Banks can avert a run if they are able to borrow from other banks and provide their customers' balances. However, if the speculation or rumors are pervasive, then banks may become unwilling to lend to each other. When this happens, it sparks even more speculation and can create a run on the entire financial system.

Bank Regulation and Deregulation

Prior to the Civil War, banks were chartered by the states and were capable of issuing their own currency. In response to the government's need for revenue to pay for the war, Congress passed the National Banking Act of 1863, which created federally chartered banks capable of issuing the new national currency and government bonds. To ensure liquidity in case of a crisis, larger banks accepted deposits from smaller banks that could be with-drawn in case the smaller banks experienced a bank run. The system was premised on the notion that a small bank run could be handled by tapping into a much larger bank's reserves. However, the system failed to recognize the possibility that a small bank run could create a contagion that would lead to a systemic run on the banks. A central bank was deemed necessary to act as a bank of last resort to stop bank panics.

FACT

Current Fed chairman Ben Bernanke is a well-known student of Federal Reserve history. His extensive knowledge of the Fed's failures during the Depression should reassure those who fear that the Fed will once again make a bad situation worse.

Bank Regulations Since the 1900s

A widespread bank panic in 1907 led congress to pass the Federal Reserve Act of 1913, which created the modern Federal Reserve System, America's version of a central bank. The Fed serves as the nation's chief bank regulator. The Federal Reserve Board of Governors regulates member banks while the Federal Reserve district banks supervise and enforce the board's regulations. Unfortunately, the Fed was put to the test during the Great Depression where it failed to provide the necessary liquidity to stem another systemic bank panic. Critics of the Fed place much of the blame for the severity of the Depression on the Fed. The general consensus is that the Fed restricted the flow of credit when it should have flooded the system with inexpensive credit.

After the stock market collapse in 1929 and the ensuing financial and economic crisis, Congress passed the Glass-Steagall Act of 1933, which created the Federal Deposit Insurance Corporation (FDIC) to insure bank deposits and prevent future bank runs. Glass-Steagall also prohibited commercial banks from engaging in most investment activities. The Bank Holding Company Act prohibited banks from underwriting insurance, further reinforcing Glass-

Steagall. Bank regulations had the effect of restoring confi-dence in the industry.

A test of the banking system came in the 1980s with the savings and loan crisis. Aggressive lending by the savings and loan industry and lax underwriting led to a series of savings and loan failures. Similar to the FDIC, the Federal Savings and Loan Insurance Corporation (or FSLIC) paid depositors whose institutions had failed. The American taxpayer was ultimately the loser as billions were spent to clean up the financial mess and refund depositors.

Throughout the twentieth century, the American economy grew and industry began to increase in size and importance. Soon, local and regional firms were competing against national firms. American businesses that were national in scope were being served by a banking system that was fragmented and regional. Bank regulation kept American banks relatively small compared to banks in other countries. The banking sector effectively lobbied for deregulation in order to grow and compete at a national and even international level.

Bank Deregulation

The deregulation of banking that occurred in the late twentieth century allowed banks to operate nationwide and also allowed them to expand the level of services they provided. Eventually Glass-Steagall was repealed and banks were engaged once again in the business of speculative investment. As the walls separating traditional banks from banklike institutions came down, the seeds for another financial crisis were sowed. Today, the banking industry is in flux. A push for regulation to prevent future bank crises exists. As the line between banks and other financial institutions has blurred, the task for lawmakers is to create a regulatory framework that encompasses all banklike activities. History will show whether or not they were successful.

CHAPTER 6: Supply and Demand

You are watching the news when someone says, "Higher gas prices should lower demand." How do you evaluate a statement like that? This guy is on television, so he must know what he is talking about, right? Don't be so sure. Many intelligent individuals throw around economic arguments using economic terminology and may even sound convincing, but there is a lot of bad economics going around! In order to make informed decisions and not be easily swayed, it is important to understand economics. To have a good understanding of economics, you must have a grasp of supply and demand. The concepts are central to even the most complex of economic arguments, yet they are easily understood.

Markets

Markets are places that bring together buyers and sellers. However, markets do not have to be physical places. Markets exist whenever and wherever buyers and sellers interact, be it a physical location, via mail, or over the Internet. Several conditions must be met in order for markets to function efficiently. Typical conditions for an efficient market include a large number of buyers and sellers acting independently according to their own self-interest, perfect information about what is being traded, and freedom of entry and exit to and from the market.

FACT

> Economists disagree over the efficiency of markets. Some argue that the market price effectively captures all of the available information about the product. Others argue that prices do not reflect all available information. They argue that this information asymmetry undermines market efficiency.

A large number of participants in the market ensures that no one buyer or seller has too much influence over the price or the amount traded. It is obvious that if there is a single seller or single buyer, they will be able to exercise considerable influence over prices. For example, Wal-Mart has what is called **monopsony power** over several producers. Because Wal-Mart is the sole retailer for these producers, it is able use that power to influence the price it pays. In competitive markets, no one producer or consumer exercises that level of influence.

Perfect information implies that both buyer and seller have complete access to the costs of production, perfect knowledge of the product, and no opportunities exist for **arbitrage**, which is buying low in one place and selling high in another. Contrast that condition with the experience of buying a car. Chances are the seller has the bulk of information about the cost and specifications of the vehicle, whereas you deal with limited information at best in making your purchase decision.

Freedom of entry and exit into the market also increases the market's efficiency by allowing the maximum number of buyers and sellers to participate. Licensing requirements are an example of a barrier to entry. By requiring licenses to sell or produce goods and services, the government limits the potential number of sellers, resulting in less competition and higher prices.

Consumer Behavior

Economists are always making assumptions about people's behavior. One assumption that seems to make sense is that people try to make themselves as happy as possible. In economics-speak, this is called **utility maximization**. When consumers buy goods and services, they are doing so to maximize their utility. The ability of consumers to maximize utility is constrained by the amount they have to spend; this is called a **budget constraint**. When it comes to consumer behavior, economists argue that consumers seek to maximize their utility subject to their budget constraint.

QUESTION

What does utility have to do with happiness?

If you have ever studied the work of the philosophers John Stuart Mill or Jeremy Bentham, then you should be familiar with the concept of utility. Bentham viewed people as lovers of pleasure and haters of pain. In his theory, people try to maximize utility (pleasure) and minimize disutility (pain).

In Chapter 1, you learned that economic decisions are made at the margin. When studying consumer behavior, it is important to focus on these marginal decisions. Suppose you are faced with a bowl of your favorite candy. Each time you benefit from consuming a piece, economists say that you are experiencing marginal utility, but you do not care what economists say because you are eating candy and getting fat and happy. You might describe the feeling you are having as getting happy points. The more candy you eat, the more happy points you get. You are maximizing your utility.

If you have ever overindulged on candy, cookies, brownies, or ice cream, you know that at first the eating was enjoyable, but after a few (or a dozen), you were not quite as happy as when you started. Economists refer to this phenomenon as **diminishing marginal utility**. In other words, each piece of candy you eat gives you fewer happy points than the prior piece. Diminishing marginal utility is a useful concept. Throw the term around the next time you are at a dinner party and see how it goes.

Diminishing marginal utility helps to explain many of your behaviors. If one peppermint makes your breath smell fresh, then why not eat fifty? Answer: diminishing marginal utility. If exercising for one hour is good for your health, then why not exercise for 24 straight hours? Again, the answer is diminishing marginal utility. If a drop of perfume makes you smell nice, why not wear the whole bottle? You guessed it — diminishing marginal utility.

Demand

If you have ever witnessed an auction, then you might have noticed that there are many more low bids for an item than high bids. That is, people are more willing and able to pay a low price for an item than to pay a high price. This willingness and ability to buy something is referred to as **demand**. The fact that more people are willing to buy at lower prices than at higher prices is called the law of demand.

Reasons for the Law of Demand

Three reasons explain why the **law of demand** exists: diminishing marginal utility, income effect, and substitution effect. The reason that diminishing marginal utility is an explanation of the law of demand is easy to understand. As you consume more and more of an item, each successive unit provides less utility, or happy points, than the previous unit. As a result, the only way that you will buy more of an item is if the price is lower. You consume until the marginal benefit (utility) equals the marginal cost (price). Assume that your favorite doughnut stand is offering a "buy two and get the third for half-price" deal. On a normal day, you would just buy two doughnuts because a third is not worth it to you due to diminishing marginal utility. But in the case of the deal they are offering, if they lower the price or marginal cost to the point where it is equal or less than the marginal utility or marginal benefit, it makes sense for you to purchase the third doughnut.

Income effect is based on your budget constraint. As the price of a good drops, your purchasing power increases. As the price increases, your purchasing power falls. Income effect explains the logic behind discounts and sale prices. When goods go on sale at a lower price, your limited income is able to purchase more, so that is what you do.

Substitution effect is another explanation for the law of demand. The substitution effect says that you substitute relatively less expensive items for relatively more expensive items. For example, imagine you are at the grocery store to buy food for five days' worth of meals — three chicken dinners and two dinners with beef. If the store happens to have beef on sale, you might substitute one day's chicken with beef. So what happened? The law of demand happened. Beef prices were relatively lower and you bought more beef.

FACT

Purchasing power and prices are inverses of each other. As prices rise, purchasing power falls and as prices fall, purchasing power rises. Economists often refer to the concepts of **real income** and **nominal income**. Real income is based on your purchasing power while nominal income is the dollar amount earned. When considering your living standard, it is real income that counts.

Elasticity of Demand

Think about all of the things that you buy in a year. You might purchase goods and services as diverse as chewing gum, emergency room visits, and cars. Sometimes you are very sensitive to the price and at other times you are not. For example, you're more likely to shop around for a good price on a car than on an emergency room visit. Economists refer to this sensitivity to price as **elasticity of demand**. When you can delay the purchase of a good, if it has many close substitutes, or if it takes a large percentage of your income, demand is typically price sensitive or elastic. If, however, the purchase must be made immediately, no close substitutes exist, or the purchase does not take a significant percentage of income, demand is price insensitive or inelastic.

Compare an emergency appendectomy with a facelift. Both are surgeries, but consumer demand for these surgeries is quite different. Acute appendicitis does not wait for you to shop for the best price on surgery. Price is probably the last thing on your mind when experiencing this illness. Your demand for the surgery is inelastic. Facelifts are an entirely different matter. First, the purchase can be delayed. Next, a facelift has available substitutes, like Botox and collagen injection. Finally, because facelifts are optional and not covered by most health insurance plans, they tend to take a large percentage of a person's income. The result of this combination of factors is that, for most, facelift demand is elastic.

Changes in Demand

Several things affect the demand for a good or service, but the one thing that will not affect the demand is the price. As the price of a good increases, you are willing to buy less, and as the price drops, you are willing to buy more. Your demand has not changed in response to changes in price, just the amount or quantity you are willing to buy. Consumer tastes, related prices, income, the number of buyers, and expected prices are factors that will affect the demand for goods and services.

ESSENTIAL

Determining whether goods are substitutes or complements is not a matter of opinion. Economists calculate **cross-price elasticity** to determine whether goods are complements or substitutes by dividing the percentage change in the quantity demanded of one good by the percentage change in the price of another good. If cross-price elasticity is less than zero, goods are complements, and if cross-price elasticity is greater than zero, goods are substitutes.

Consumer tastes for goods and services are subject to change and when these changes happen, demand shifts. A few years back, there was a recall of fresh leaf spinach because of *E. coli* contamination. This contamination effectively reduced demand for fresh spinach. Advertising can affect the taste for goods and services as well. The ShamWow is a case in point. The ads for this product were on several channels on TV. After watching that ad for the hundredth time many people decided that they *had* to have one. The increase in ShamWow sales was not in response to some change in the price of the product but was caused by a change in consumers' tastes due to advertising.

The price of related goods can affect the demand for a good as well. Related goods are classified as either complements or substitutes. **Complements** are goods used in conjunction with other goods, and **substitutes** are goods used in lieu of each other. Movie tickets and Junior Mints are complements. As the price of movie tickets rises, people are less willing to buy them and go to the movie. Therefore, there is less demand for Junior Mints. If ticket prices fall, the opposite occurs. Air travel and bus travel are substitutes. As the price of airline tickets falls, demand for bus tickets decreases, and as the price of airline tickets rises, demand for bus tickets increases.

Changes in consumers' income will lead to changes in demand. If income and demand move in the same direction, you are dealing with a **normal** good. If income and demand move in opposite directions, the good is considered **inferior**. If organic milk is a normal good and powdered milk is an inferior good, what effect will an increase in consumers' income have on demand for the two?

Demand for organic milk increases with increased income, and demand for powdered milk decreases with increased income.

The number of buyers is directly related to the demand for goods and services. As the number of buyers increases, so will demand, and vice versa. This obvious relationship should be considered before investing in the Russian housing market. The population there is shrinking.

QUESTION

How do you know whether a good is normal or inferior?

Economists calculate **income elasticity of demand** to answer the question. By dividing the percent change in the quantity demanded of a good by the percentage change in income, economists are able to classify goods as either normal or inferior. Income elasticities greater than one indicate **normal luxuries**, income elasticities between one and zero indicate **normal necessities**, and income elasticities less than zero indicate inferior goods.

Expected prices can have a direct influence over the demand for goods and services. If investors believe that a stock's price will increase in the future, demand for the stock increases. Likewise, if the price of a stock is expected to fall, the demand for the stock will decrease.

Supply

Supply is to producer as demand is to consumer. Supply is what producers do. Supply reflects producers' changing willingness and ability to make or sell at the various prices that occur in the market. If you were selling cookies or crude oil, which would entice you to produce more, a low price or a high price? If you said low price, you would quickly find yourself broke. However, if you said high price, you just might have a chance to make a profit. The **law of supply** states that producers are able and willing to sell more as the price increases. The reason for the law of supply is the simple fact that as production increases, so do the marginal costs. As rational, self-interested individuals, suppliers are only willing to produce if they are able to cover their cost.

Elasticity of Supply

Elasticity of supply is the producers' sensitivity to changes in price on the quantity they are willing to produce. The key factor in supply elasticity is the amount of time it takes to produce the good or service. If producers can respond to price changes rapidly, supply is relatively elastic. However, if producers need considerable time to respond to changes in the market price of their product, supply is relatively inelastic. Compare corn tortillas and wine. Corn tortillas are easily produced with readily available materials. If the market price of corn tortillas were to suddenly increase, producers would have little difficulty in producing more tortillas in response to the price change. Now, if the market price of pinot noir were to suddenly increase, wine makers would have much more difficulty responding to the price change. Vines take years to develop, grapes take time to ripen, and wine needs time to age. All of these factors give wine a relatively inelastic supply.

FACT

Elasticity of supply and demand is important in determining the **incidence of taxation**, or who bears the burden of a tax on goods and services. Even though the legal incidence of a tax might fall on a producer, the economic incidence of the same tax might fall on the consumer.

Changes in Supply

As price changes, producers are willing to produce more or less. Price affects the quantity producers supply, but it does not affect supply. For example, the supply of coffee is influenced by weather, land prices, other coffee producers, coffee futures, cocoa profits, and subsidies to coffee producers. The one thing that does not influence the supply of coffee is the current price of coffee. This often causes confusion, but it need not. Understand that supply refers to producers' willingness to produce various amounts at various prices, and not to some fixed quantity. Supply is influenced by nature, the price of inputs, competition, expected prices, related profits, and government.

Nature plays a big part in determining the supply of coffee. Rain, sunshine, temperature, and disease are obvious examples of variables in nature that will affect the coffee harvest. Excellent weather conditions often lead to large increases in supply, and poor weather leads to the opposite.

ESSENTIAL

Economists recognize that expectations of the future play a large role in determining the economy of today. As time has progressed and policymakers have become more sophisticated, much effort has been placed in setting expectations for the markets. Instead of reacting, policymakers must now contend with setting future expectations when they make policy decisions about the economy.

Input or resource prices have a direct influence on producers' supply decisions. Land, seed, fertilizer, pesticide, harvesting equipment, labor, and storage are just a few of the costs that coffee producers face. Supply decreases as those costs rise, making growers less able to produce at each and every market price. Supply increases when the cost of production falls.

The presence of more or less competition causes increases or decreases in supply. As the popularity of coffee has risen, more and more producers have entered the market. The introduction of more competition increased the quantity of coffee supplied at each market price.

Expectations of future price increases tend to decrease supply, but expectations of future price decreases have the opposite effect. If producers expect higher prices in the future, they will be less willing to supply in the present. Coffee producers might withhold production in order to sell when prices are higher. If prices are expected to move lower in the future, producers have an incentive to sell more in the present.

The profitability of related goods and services also affects the supply of a good like coffee. For example, coffee-growing land is also favorable for growing cocoa. If the profits are greater in the cocoa market than in the coffee market, over time more land will be pulled from coffee production and put into cocoa production. Likewise, if profits in the coffee market are greater, eventually, more land will be put into coffee production at the expense of cocoa production.

Government policies can also affect supply. Government can tax, subsidize, or regulate production, and this will affect supply. If Brazil wants to reduce the local production of coffee in order to restore forests, the Brazilian government can tax coffee production. This would increase the cost of production and reduce the supply. If government wants to encourage production and increase supply, it can subsidize producers, that is, pay them to produce. Vietnam might subsidize coffee production in its highlands in order to increase the supply of this valuable export commodity. Regulation often has the effect of limiting supply. If Vietnam wanted to preserve its highland rainforests, it might make rules or regulations that effectively limit the ability of coffee growers to produce.

Technology and the availability of physical capital are key determinants of supply. Technological innovation has allowed producers in many different industries to increase the quantity of goods that they are willing and able to produce at each and every price. Increases in the amount of physical capital available relative to labor also help firms to increase output. Economists refer to this phenomenon as **capital deepening**. As capital deepening increases for a firm, so does supply.

A Price Is Born

When supply meets demand, something interesting happens. A **price** is born. In an efficient market, prices are a function of the supply and demand for the good or service. Instead of central planners, government officials, or oligarchs dictating artificial prices or rationing who gets what, the market relies on the impersonal forces of supply and demand to determine prices and to serve the rationing function. The pitting of consumers trying to maximize their utility against producers trying to maximize their profits is what determines the price of goods in the market and also the quantity that is bought and sold.

QUESTION

Which is better, allowing the price system to ration, or allowing a select group in society to determine who gets what?

During World War II, the United States government instituted rationing in an effort to divert resources toward the war effort. Even if you had the money to buy what you wanted, rationing required that you also have the ration coupon. No ration coupon, no rations. Fortunately, America returned to the market system in 1946. The price system is preferred because it allows people to get what they want, not just what they need.

Supply and demand ration goods and services efficiently and fairly. Prices are efficient because they are understood by most participants in the market. If you give a child five dollars and send her into a candy shop, she could figure out what she can afford without having to ask anyone for help. A price conveys much information. The price of a good communicates to consumers whether or not to purchase and to the producer whether or not to produce it. Prices are fair because they are neutral; they favor neither buyer nor seller.

Finding Equilibrium

A market is said to be in **equilibrium** when at the prevailing price there is neither a surplus nor shortage of the good or service. When this condition is present, then the price is called the equilibrium or **market clearing price**. Market equilibrium is the most desirable outcome because it allows for consumers to maximize utility while also allowing producers to maximize profits.

There are times when the market is not in equilibrium. Sometimes, the market price is greater than the equilibrium price. When this happens, a **surplus** results. The amount producers supply is greater than the amount consumers demand. If you have ever walked past a clearance rack full of sweaters and wondered to yourself, "Who would wear that?", you are not alone. Countless others had walked past those now-surplus sweaters before they were placed on the clearance rack. They walked past because the marginal cost of the sweater to the consumer was greater than the marginal utility. The purpose of the clearance rack is to offer these sweaters at a price low enough to induce some hapless, utility-maximizing individual to purchase them.

If the market price is too low, then a **shortage** might result. Shortages occur when the quantity demanded is greater than the quantity supplied. When shortages occur in the market, buyers compete against each other to purchase an item and bid up the price until equilibrium is reached. Auctions take advantage of this phenomenon, and the consumer who wants the good the most gets it. How do you know he wanted it the most? He offered the most money. Prices are fair, efficient, and effective at rationing most goods and services.

Changes in Demand and Supply

Change in either demand or supply will cause change in both price and quantity. Suppose people started seeing this headline everywhere: "Medical researchers discover that drinking coffee has immediate health benefits." What do you expect would happen to the price and quantity of coffee that is exchanged in the market? The news might alter consumer tastes for coffee and lead to an increase in the demand for the drink. As demand increases, the quantity that consumers are willing and able to purchase at every price increases. Because coffee is relatively scarce and its producers face increasing marginal cost, the equilibrium price and quantity of coffee will rise in response to the increased demand.

What if Canada blocked all oil exports to the United States? How would that affect the market for gasoline? Canada is the largest exporter of oil to the United States, so this would definitely disrupt gasoline production. Decreased oil supplies would lead to a higher price and a lower quantity of oil. Because oil is the primary input of gasoline, the effect of higher oil prices would be to raise the cost of gasoline production. As the cost of production rises, producers supply less gasoline at every price in the market. Consumers, dutifully obeying the law of demand, are less willing to purchase gasoline as the price rises. The end result of the Canadian oil embargo would be to raise gasoline prices and reduce the quantity of gasoline sold.

ALERT

It is easy to fall into the trap of attributing prices to "them." "They" charge so much! "They" should lower prices! In a market, prices are determined by consumers and producers. If "you" would stop buying so much, "they" would charge less.

How does an economic recession affect the price of lobster? As unemployment increases, household incomes decrease. This decrease in income tends to reduce the demand for normal goods such as lobster, so at each and every price, consumers are not willing to buy as much as before. At these lower prices, many producers are unable to profit, and as a result the price and quantity of lobster sold in the market decreases.

Technological innovation often leads to greater productivity and lower production costs. Consider Henry Ford's application of the assembly line to automobile production. Ford was able to offer more cars for sale at each price in the market because of this innovation. This, combined with consumers' willingness to purchase more at lower prices, or the law of demand, resulted in the equilibrium price of cars decreasing while the equilibrium quantity increased.

Expected prices affect both supply and demand. If both consumers and producers expect home prices to decrease in the future, consumer demand decreases and producer supply increases. Consumers will now wait for the lower prices to materialize and hold off on making immediate purchases. Producers or sellers will seek to unload their inventory now while prices are still high. The combined result of this expected price decrease is a price decrease. Self-fulfilling prophecy is at work. What about quantity? Without knowing the actual values, you are unable to accurately predict what will happen to quantity. Supply and demand as presented allows you to estimate whether prices and quantities are going up or down, but without actual numbers, it is difficult to predict outcomes when supply and demand simultaneously change.

Markets Talk

As you can see, a variety of factors affect supply and demand, which in turn affect price and quantity. Changes in the market for one good will create changes in the market for another good. This happens as price changes are communicated across markets. This phenomenon should be considered when policymakers attempt to influence markets because unintended consequences can result.

How might driving an SUV contribute to starvation in Southeast Asia? Several years ago gas prices suddenly began to climb. As a result, there was considerable political pressure to alleviate the squeeze placed on the pocketbook of many Americans. Instead of driving less or commuting, many wanted to continue their lifestyle of driving an inefficient vehicle without having to pay higher prices. According to Thomas Sowell of the Hoover Institute, politicians and many people are fond of ignoring the aphorism "there is no such thing as a free lunch." So here is what happened.

As gas prices increased, demand for alternative fuels increased. This increase in demand for alternative fuels was popular among corn growers who had a product called ethanol. In order to provide ethanol at lower cost, corn growers lobbied Congress for greater subsidies. This resulted in more land being placed into corn production at the expense of other crops, namely wheat. As wheat supplies decreased and wheat prices rose, the price of substitute crop, rice, also rose because there was now more demand for rice. This led to the price of rice increasing to the point where people in Southeast Asia were unable to afford their basic staple. Starvation quickly ensued. Markets talk to each other. No one intended for starvation to occur, but when people ignore scarcity, unintended consequences can and do occur.

CHAPTER 7: Competitive Markets

Do you remember the first time you learned about atoms in science class? The teacher probably drew a sketch on the chalkboard that looked like a model of the solar system: a big nucleus in the middle orbited by a tiny electron. Later you probably learned that atoms do not actually look like the drawing on the board, but the model your teacher showed you helped you to understand atoms. In economics, when studying markets, you begin by learning something that is somewhat unrealistic, but a simple model of perfect competition will help you to understand real-world conditions.

Conditions

Certain conditions are necessary for the functioning of an efficient market: a large number of buyers and sellers each acting independently according to their own self-interest, perfect information about what is being traded, and freedom of entry and exit to and from the market. Add to this list that firms deal in identical products and that they are "price takers," and you now have **perfect competition**.

Identical products mean that there are no real differences in the output of firms. They are all making and selling the same stuff. Think of things like wheat, corn, rice, barley, and whatever else goes into making breakfast cereal. Wheat grown by one farmer is not significantly different from wheat grown by another farmer.

QUESTION

Why are perfectly competitive markets preferable to other types of market?

Perfectly competitive markets are what economists call **allocatively efficient**. Consumers get the most benefit at the lowest price without creating any loss for producers. Perfect competition is also productively efficient because in the long run, firms produce at the lowest total cost per unit.

Economists refer to firms as "pricetakers" when a firm does not set the price of its output but instead sells its output at the market price. Remember, one outcome when markets have many different small buyers and sellers is that none are able to influence the price of the product.

Accounting Versus Economics

An accountant asked an economist why she had chosen a career in economics over accounting. The economist replied, "I'm good with numbers, but I don't have enough personality to be an accountant." Personality differences aside, a key distinction between economics and accounting is in determining total cost and profit. To the accountant, total costs are the sum of all of the explicit fixed and variable costs of production. In addition, profits are equal to total revenue minus total cost. To the economist, however, total cost is equal to all of the explicit fixed and variable costs plus opportunity cost. Profits to the economist are equal to total revenue minus total cost including opportunity cost.

Imagine that you are a teacher earning $5,000 a month and decide to quit your job and start selling snow cones instead. You buy a freezer cart that you can wheel around, order all of the supplies you need, and pay the required licensing fees. Assume your total cost equals $2,000. So you get out there and start hustling snow cones, and you're actually good at it. At the end of the month, you calculate that you have earned $6,000 in total revenue. What are your accounting profits? $6,000 in total revenue − $2,000 in total cost = $4,000 in accounting profit. What are your economic profits? $6,000 in total revenue − ($2,000 in explicit cost + $5,000 in opportunity cost) = -$1,000 economic loss. The opportunity cost is what you could have been earning as a teacher.

FACT

Many people confuse the concepts of revenue and profit. **Revenue** is all of the income a business earns. For a firm selling a single type of product at one price, revenue is equal to the quantity sold multiplied by the price. **Profit**, on the other hand, is the income a company has left over after covering all of its costs. Revenue − Cost = Profit

Economic profits in an industry are important because they provide firms in other industries with an incentive to employ their land, labor, capital, and entrepreneurial ability in the economically profitable industry. Economic profits draw resources to their most efficient use. In the long run, competition eliminates economic profits. Industry is most efficient when economic profits are equal to zero. At zero economic profit, there is no incentive for existing firms to leave or for new firms to enter the market. In the example above, the $1,000 economic loss is a signal for you to leave the snow cone industry because your resources could be put to better use in another industry.

The Production Function

Both microeconomics and macroeconomics make distinctions between the short run and the long run. These distinctions have very little to do with some fixed period of time but rather are based on the ability of firms to make changes in their inputs. The **short run** is defined as the period of time in which firms are able to vary *only one* of the inputs to production, usually labor. The **long run** is the period in which firms are able to vary all the inputs in the production process. If you operate a restaurant, in the short run you can only add or subtract workers to adjust the level of production. If your place is busy, you schedule or call up more of your workers. If business is slow, you send the employees home. The long run is the period in which you are able to expand the kitchen or add new equipment. So in response to an increase in business activity, in the short run you can schedule more workers, but in the long run you can make the restaurant bigger.

A firm's short-run production decisions are based on the firm's **production function**. A production function shows how a firm's output changes as it makes changes to a single input, like labor. The production function is divided into three distinct stages based on what is happening to the firm's output or product.

The Stages of Production Function

The first stage of a production function occurs when firms experience **increasing returns**. This means that as a firm adds workers, each additional worker contributes more to output than the previous worker. The additional contribution to output from each worker is referred to as **marginal product**. So in the stage of increasing returns, both output and marginal product are increasing.

The second stage of production is called **diminishing returns**. In this stage, as the firm increases the number of workers, output still increases, but the additional contribution of each worker decreases. Finally, the firm experiences the third stage of production called **negative returns**. In this stage, as firms add workers, both output and marginal product decrease.

To illustrate the production function, picture a restaurant on a normal business day. Assume that you are the manager responsible for scheduling workers, but you have no experience and are a little slow at figuring things out. At 6 A.M. the first customer arrives for breakfast. You immediately call up one of your employees and have her rush to work. Employee 1 is able to prepare the food, serve it, and then act as a cashier for the transaction. Later, as more customers arrive and begin demanding service, it becomes obvious to you that more help is needed. Taking out your cell phone, you call up Employees 2 and 3 and order them to work. As they settle in and begin working, a division of labor develops, which increases their individual productivity and the total productivity of the restaurant.

Witnessing the marvelous outcome, you conclude that more is always better and decide to call up Employees 4 through 7. As they begin to work, you notice that the restaurant is able to serve more customers, but the earlier gains in productivity are beginning to diminish. You chalk up these diminishing gains as a fluke, and in order to break through this impasse you bring in Employees 8 through 256. Pretty soon, the restaurant's kitchen is full of employees, with every worker pretty much immobilized like a sardine in a can. The customers are now outraged at the extremely slow service and a little freaked out by your lack of management ability. As a matter of fact, with 256 workers, you are unable to produce anything. This is referred to as negative returns.

Cost, Cost, Cost

Businesses or firms strive to earn the most profits possible. They do this by trying to increase revenue and decrease cost. In a competitive environment, there is not much firms can do to increase revenue. They lack pricing power. Firms do have the ability to control cost, so in order to maximize their profits they try to produce at the lowest cost possible.

Economists break down costs into different categories, the first of which are **fixed costs** or **overhead**. A firm's fixed costs are those costs that don't change regardless of the level of production. Rent, property tax, management salaries, and depreciation are examples. Whether a factory is running at full capacity or is idle, the overhead remains the same. Firms also face **variable costs**. Variable costs change with the level of a firm's output. Utilities, hourly wages, and per unit taxes are representative of variable costs. As a firm's production increases, so do its variable costs. **Total cost** is the sum of a firm's fixed and variable costs.

The marginal cost of production is of special interest to economists. Marginal cost is the change in total cost for each unit produced. Think of marginal cost as the additional cost of producing one more item. For each additional unit of output a firm produces, it incurs more variable cost and hence more total cost. This means that its marginal cost increases as well. For example, each Big Mac costs more to produce than the previous Big Mac because McDonald's had to pay for more ingredients and pay its workers more for the extra time it took to produce the additional Big Mac. Firms like McDonald's maximize their profits when they produce at the point where marginal cost equals **marginal revenue**. In other words, if a firm wants to make the largest profits it can, it will produce up to the point where the additional cost of producing one more item is the same as the additional revenue earned by producing one more item.

ESSENTIAL

In the long run, all costs are variable. Over time, firms are able to add or subtract capital, renegotiate rent, and alter management salaries. The distinction between fixed and variable costs disappears with the passage of time.

Perfect Competition in the Short Run

For an industry, the short run is the period of time in which firms are unable to enter or exit the market because they are only able to vary their labor and not their fixed capital. In the short run, it is possible for firms in a perfectly competitive industry to earn economic profits or even operate at a loss as supply and demand for the entire industry's output changes.

Assume that the glazed doughnut industry is perfectly competitive. Imagine that scientists working in New Zealand discover that glazed doughnuts, when consumed with coffee, are extremely beneficial to consumers' health. As a result of this great news, the demand for doughnuts increases. This results in a new, higher equilibrium price. Remember that the market price represents the firm's marginal revenue, so for firms in the doughnut industry, their total revenue has increased by more than their total economic cost. This means that glazed doughnut firms are earning economic profits.

Six months later, scientists in California reveal that the earlier New Zealand doughnut research was flawed and that, in fact, consuming large amounts of glazed doughnuts with coffee might pose a risk to consumers' health. First there is denial, then consumers slowly awake to the reality that they are fifty pounds heavier and finding it difficult to sleep. At this point, the demand for glazed doughnuts decreases below its original equilibrium. For many firms in the doughnut industry, this decrease in the market price means that they are now producing at a short-run loss because their total revenue is less than their total cost of production.

QUESTION

At what point does a firm shut down for good?

Firms will operate, even at a loss, as long as they are able to cover their variable cost. At the point where revenues are less than variable cost, firms must shut down because they are unable to pay workers and keep the lights on.

Perfect Competition in the Long Run

In the long run, firms are able to enter and exit the market. New firms enter the market in response to the presence of economic profits, and old firms exit the market in response to losses. In light of the New Zealand research, economic profits in the glazed doughnut industry would attract new firms to the industry. As new firms enter, competition increases, which means that the industry supply of glazed doughnuts increases. The increased supply reduces the equilibrium price of doughnuts and economic profits disappear.

Considering the California research, in the face of economi[c losses,] some firms will reach the shutdown point and withdraw [from the] industry. This reduces competition and decreases the [total] supply of glazed doughnuts. Decreased supply increases [the] equilibrium price in the market. In the end, fewer firms remain as the industry returns to its long-run equilibrium with zero economic profits.

From Competition to Imperfect Competition: A Continuum

Perfect competition does not really exist. It is unlikely that you will find an industry in which all of the conditions for perfect competition are met. However, perfect competition provides an example with which to compare the market structures that do exist. Although it isn't real, it provides a nice frame of reference.

FACT

> The continuum of market structures can be seen as an evolution of markets. Firms may begin in a very competitive market and over time become monopolists. The late nineteenth and early twentieth centuries witnessed the rise of the trusts from once-competitive industries. Some even called for an end to "ruinous competition."

Most firms face barriers to entry, either in terms of cost or government requirements. Firms rarely deal in identical products, as they invest heavily in differentiating themselves from the competition. This ability to differentiate gives firms some ability to affect prices. In mature markets like the United States, firms tend to be large and not necessarily independent. Finally, access to information is not equally shared, and as a result, the condition of perfect information does not exist either. What you are left with is not perfect competition but **imperfect competition**.

Economists classify markets according to their level of competition. On one end of the spectrum lie perfectly competitive, albeit fictional, markets. On the other end of the continuum lies monopoly. In between you will find the market structures that are most familiar: monopolistic competition and oligopoly.

Monopolistic Competition

Monopolistic competition is a market structure very similar to perfect competition. There are many buyers and sellers, barriers to entry are minimal or at least equal for all firms, and information is readily available. However, in monopolistic competition, firms do not offer identical products but differentiate their products from their competition. **Product differentiation** is the process by which producers are able to convince consumers that their particular product is different from other producers.

The industry that should come to mind when you think of monopolistic competition is fast-food. The fast-food industry has many different producers competing for the dollars of many different consumers. All are welcome to start a fast-food restaurant as long as they pay the required licenses. Most producers have a good idea of what they are getting into and customers tend to understand the products quite well. Why is fast food monopolistically competitive? Product differentiation. Each firm offers a different menu. Taco Bell, Chick-Fil-A, McDonald's, and Subway all compete against each other in the fast-food market while providing customers with a variety of choices. Product differentiation is one of the reasons that new entrants are able to survive in this cutthroat industry. If you are different enough, then you might have a chance.

QUESTION

Can monopolistically competitive firms maintain economic profits in the long run?

No. Over time, the presence of competition will eventually erode the monopolistically competitive firm's profits. The end result is an industry with excess capacity, high cost, and no economic profits.

Next time you are in the produce section of the grocery store, take a look at how many varieties of apples are available. Do you remember the days when apples were either red or green? Today there are Red Delicious, Pink Lady, Granny Smith, Gala, Fuji, and Honeycrisp, to name a few. In addition, there are small, medium, and large. There are organic and pesticide-treated. Some are sold individually and some are packaged. Apple producers have differentiated their product. Why? Remember how competitive firms were unable to influence price? When producers differentiate their product and are successful, they are able to charge a higher price than their competition.

The problem with product differentiation is that it becomes a never-ending process. Firms must continually find ways to differentiate. This explains why firms will spend large sums of money on advertising. Much of the advertising is not so much an attempt to gain new customers as it is an effort to build brand loyalty. However, firms have limited resources, so engaging in product differentiation through advertising means that resources used in advertising are no longer available for production. As a result, industries that are monopolistically competitive do not produce as much output as they could if they were perfectly competitive. Consumers are missing out on what could have been. Not to worry, it seems that consumers have a strong preference for the variety that monopolistic competition brings, and so the benefits may at least equal the costs.

CHAPTER 8: Imperfectly Competitive Markets

Many of the companies with which you are most familiar do not exist in highly competitive markets. The major automobile manufacturers, airlines, telecommunications companies, food producers, and discount retailers compete in an oligopolistic market structure. At a local level, many of the utilities you use are monopolies. Oligopoly and monopoly are common market structures in the United States. To know how the American economy works, you need a good understanding of these imperfectly competitive market structures.

Oligopoly — Few Dominate

Oligopoly describes a market where a few large producers dominate. Unlike competitive and monopolistically competitive markets, oligopolistic firms have more pricing power. In addition, oligopoly is characterized by considerable barriers to entry because of the sheer scale of the firms. Oligopoly is often the result of once-competitive markets maturing. As monopolistically competitive firms grow and merge with other firms, fewer firms result. Oligopolies are of concern to government regulators attempting to preserve and enforce competition in industries. As competition decreases, prices become higher, and productive and allocative efficiency, which benefit society, are lost.

FACT

When firms in the same stage of the production process join to form a single firm, it is called a **horizontal merger**. Chevrolet, Buick, Cadillac, and GMC form the horizontal merger better known as General Motors. **Vertical mergers** occur when firms in different stages of the production process merge. Andrew Carnegie employed this strategy in making Carnegie Steel the industrial titan of its day. He bought up firms in every step of the production process, from raw iron ore to the finished steel.

To determine whether a market meets the condition of oligopoly, economists calculate the **Herfindahl-Hirschman index** (HHI) for the market. A relatively low index number identifies a market as competitive, and a relatively high index number indicates oligopoly. The Federal Trade Commission (FTC) and the Justice Department can use the index numbers as a way to determine whether or not to approve mergers between companies in an industry. If the merger would significantly increase the HHI, then the merger would most likely be blocked because it would reduce competition.

Regulators and economists also use concentration ratios to determine if a market is oligopolistic. The more market share is dominated by a few firms, the higher the concentration ratio. For example, if the four-firm concentration ratio is 80%, then the four largest firms have 80% of the market share. According to the 2002 census, the four-firm concentration ratio for the vacuum cleaner industry was 78%, and the eight-firm concentration ratio was 96.1%. It is safe to say that the vacuum cleaner industry is an example of oligopoly. By way of comparison, a perfectly competitive industry would have a four-firm concentration ratio of about 0%, and an industry dominated by a monopoly would have a one-firm concentration ratio of 100%.

QUESTION

What defines an industry for the purpose of determining concentration ratios?

It depends. Today the definition of industry is becoming blurred. In the past, the newspaper industry was an industry. Today it is part of a larger industry known as the media. Even though the local paper may have 100% of the local newspaper market share, other forms of media reduce its effective share in the overall industry.

The large market share that oligopolists enjoy shapes the way they view the market. Unlike firms in more competitive market structures that behave independently of each other, the oligopolists have an interdependent relationship with each other. Because the oligopolists control so much market share, their individual decisions have considerable impacts on market prices. Knowing this to be the case, the oligopolist tends to be more aware of the competition and takes this into account when making production and pricing decisions.

Collusion and Cartels

One of the blessings of competition is that it leads to lower prices for consumers. For the producer, however, this blessing is a curse. Low prices often mean low profits. Given a choice between competition and cooperation, profit-maximizing firms would more often than not prefer cooperation. Regardless of what you learned in kindergarten, you do not want the businesses you buy from to cooperate. You want them to compete. Adam Smith, the father of modern capitalism, warned that nothing beneficial comes from the heads of business getting together.

In the United States, firms are forbidden from cooperating to set prices or production. The abuses of the late nineteenth and early twentieth century trusts were the impetus for the "trust-busting" of President Theodore Roosevelt. With the Sherman Antitrust Act and later the Clayton Antitrust Act, the government prohibited outright **collusion** and other business practices that reduced competition.

FACT

Prior to OPEC, world oil prices were mainly under the control of the Texas Railroad Commission. With the rise of OPEC came a shift in power from U.S. producers to the oil states of the Middle East.

Even though it violates the law, businesses from time to time will collude in order to set prices. Colluding firms can divide up the market in a way that is beneficial for them. The firms avoid competition, set higher prices, and reduce their operating costs. Because collusion is illegal and punishable by fine and prison, executives at firms are reluctant to engage in the practice. The meetings of business leaders are almost always in the presence of attorneys in order to avoid the accusation of collusion.

Forming Cartels

Businesses that collude may form cartels. A **cartel** is a group of businesses that effectively function as a single producer or monopoly able to charge whatever price the market will bear. Probably the best-known modern cartel is the Organization of the Petroleum Exporting Countries, or OPEC. OPEC is made up of thirteen oil-exporting countries and is thus not subject to the antitrust laws of the United States. OPEC seeks to maintain high oil prices and profits for their members by restricting output. Each member of the cartel agrees to a production quota that will eventually reduce overall output and increase prices. OPEC is bad news for anyone that enjoys cheap gasoline.

Fortunately for consumers, cartels have an Achilles heel. The individual members of a cartel have an incentive to cheat on their agreement. Cartels go through periods of cooperation and competition. When prices and profits are low, the members of the cartel have an incentive to cooperate and limit production. It is the cartel's success that brings the incentive to cheat. If the cartel is successful, the market price of the commodity will rise. Individual members driven by their own self-interest will have an incentive, the law of supply, to ever-so-slightly exceed their production quota and sell the excess at the now higher price. The problem is that all members have this incentive and the result is that eventually prices will fall as they collectively cheat on the production quota. Cartels must find ways to discourage cheating. Drug cartels use assassination and kidnapping, but OPEC uses something a little more civilized. The single largest producer in the cartel is Saudi Arabia. Saudi Arabia also has the lowest cost of production. If a member or members cheat on the cartel, then Saudi Arabia can discipline the group by unleashing its vast oil reserves, undercutting other countries' prices, and still remain profitable. After a few months or even years of losses, the other countries would then have an incentive to cooperate and limit production once again.

Game Theory

Economists have discovered that game theory is useful for understanding the behavior of oligopolists. **Game theory** looks at the outcomes of decisions made when those decisions depend upon the choices of others. In other words, game theory is a study of interdependent decision-making. One game that is particularly applicable to the study of oligopoly is the prisoner's dilemma.

Two men, Adam and Karl, are picked up by the police on suspicion of burglary. The chief investigator knows that she has little evidence against the men and is counting on a confession from either one or both in order to prosecute them for burglary. Otherwise, she can only prosecute them for unlawful trespass.

Upon entering police headquarters, the men are immediately separated and taken to different rooms for interrogation. The interrogator individually informs Adam and Karl that if one confesses to the crime and implicates his partner while the other remains silent, then the one who confesses will receive a two-year jail sentence while the silent partner will likely serve a ten-year sentence in a notorious prison. If both confess, then they will likely each serve a three-year prison sentence.

What is the best strategy for Adam and Karl? If they could get together and collude, both would probably decide that it would be wise to remain silent and serve a one-year jail term for unlawful trespass. However, they are unable to collude, so they each must consider their options. Adam thinks to himself, "If I confess, then I'll either go to jail for two years or three years. If I'm silent, then I'll spend a year in jail or go to prison for ten years." Karl thinks exactly the same thing. Because they are separated and have no idea what the other is doing, they both confess in order to avoid a possible tenyear prison sentence. They both end up doing three years in jail. This logical conclusion is referred to as a **dominant strategy**.

ESSENTIAL

Organized crime enjoys much of its success in evading prosecution to a good understanding of game theory. You talk, you die. That completely changes the payoff in the prisoner's dilemma and helps to explain why prosecutors find it difficult to get confessions from members of organized crime.

Game Theory in Business

Some business decisions follow the same logic. Assume in an isolated small town there are only two gas stations and they are out of direct sight from each other. By law they are only allowed to change their price once a day. Each firm has two pricing options available to them. They can charge a high price or a low price. From past experience they know that when they both charge a high price, they both profit by $1,000. When one charges a high price and the other a low price, then the high-priced station earns $300 in profits while the low-priced station earns $1,200. When they both charge a low price then profits are $750 for each.

If given the chance to collude, which strategy would they both take? Given collusion, both would agree to set a high price for gas and each would earn daily profits of $1,000. What should the gas stations do if they are unable to collude? The thinking goes like this. "If I charge a high price, I'll either earn $1,000 or $300. If I charge a low price, I'll either earn $1200 or $750." Because the firms are out of sight from each other and have no legal way to know the other's pricing strategy, then their best course of action or dominant strategy is to set a low price, which guarantees at least $750 in profits and as much as $1200. Just like in the prisoner's dilemma, when the players do not have the ability to collude, they select a strategy that results in an outcome that is not necessarily the one that maximizes profits.

FACT

Mathematician and game theorist John Nash is the subject of the film *A Beautiful Mind*. John Nash won the Nobel Prize in economics in 1994 for his contributions to the field. A Nash equilibrium is said to exist if a player has no incentive to independently change his course of action in a game.

Unlike the prisoner's dilemma, which is a one-time game, firms compete against each other day after day. Given the chance to play the "game" over and over results in something called **tacit collusion**. By playing a game of tit for tat, the firms can eventually reach a point where they both charge a high price and maximize their profits. How does it work? Assume that on the first day, both gas stations charge a high price. Both earn profits of $1,000. On the second day, one of the stations charges a high price, but the other cheats and charges a low price to earn profits of $1,200. Predictably, the next day the other station retaliates and lowers its price, resulting in profits of $750 for each. Eventually both gas stations come to the realization that if they both set a high price and do not cheat, they both will earn higher profits in the long run. They learn that if they cheat, their additional profit for the next day will not offset the lower profits that will ensue.

Pricing Behaviors

Interdependence leads oligopolists to behave strategically. The strategic pricing behaviors that occur in oligopoly include price leadership and price wars. In addition to these pricing behaviors, oligopolies also engage in **nonprice competition**. The purpose of these price and nonprice behaviors is the same, however, and that is to maximize oligopolistic firms' profits.

Price leadership takes place when a dominant firm makes the pricing decision for the rest of the market. These decisions are often made public long before the new price goes into effect and represent a form of tacit collusion. Smaller firms in the industry will usually follow suit and match the price leader's price. Price leadership offers firms an opportunity to capture a price that is higher than would occur if the firms directly competed on price. Consumers usually fare better under price leadership than they would if the firms formed a cartel.

Price wars occur when firms break out of the price leadership model and begin undercutting one another's prices. Although it sounds bad, price wars are often advantageous to consumers because of the competitive prices created in the process. Some firms have been accused of financing price wars by raising prices in one part of their market in order to cut their price in another part of their market. A price war continues until the firms once again reach tacit collusion and return to the price leadership model.

While in the price leadership operating mode, firms compete on the basis of product differentiation as opposed to price. By emphasizing their product's differences and uniqueness, firms attempt to wrest market share from one another. As in the monopolistically competitive market, oligopolists engaging in nonprice competition will spend large sums on advertising. For example, the major

American beer brands do not compete on price, but instead rely on nonprice competition in the form of advertising in order to gain market share from one another.

Monopolies → Dominated By One

On the opposite end of the spectrum from perfect competition lies monopoly. As the name suggests, **monopoly** is a market dominated by a single seller. In the United States, monopolies are generally not allowed to exist, and every effort is made by government regulators at the FTC to prevent their creation. The reason for this prohibition is that monopolies create a serious problem for both consumers and likely competitors in the marketplace. Despite the fact that monopolies are undesirable, there are several good reasons for some to exist. Economies of scale, geography, government protection, and government mandate are primary reasons for the existence of most monopolies in the United States.

Monopoly occurs when a competitive firm eliminates all competition. Through control of key resources, mergers, and even a little help from government, once-competitive firms may find themselves in the enviable position of being a monopolist. John D. Rockefeller's Standard Oil Trust is probably the most notable American monopoly. By controlling the resource, purchasing the competition, and having political influence, Standard Oil at its height of power virtually controlled all oil production in the United States. This was good for Mr. Rockefeller as it made him the world's wealthiest man, but for consumers and possible competitors, the results of the Standard Oil Trust were high prices, inefficient production, and significant barriers to entry. Eventually, the Sherman Antitrust Act was used to breakup Standard Oil, and ever since the government has taken an active role in preventing further monopolies.

FACT

Monopolies produce less output than would occur if the market were competitive. Monopolies also charge higher prices than competitive firms. Monopolies result in what economists refer to as **deadweight loss**, which is the difference in outcomes between the inefficient monopoly and the efficient competitive market equilibrium.

Price Discrimination

Because they lack competition, monopolies can engage in price discrimination to make it difficult for other firms to do business. **Price discrimination** is the ability to charge different customers different prices for the same good or service. For example, a railroad monopolist could charge different rates to different customers for carrying the same amount of freight. Today, thanks to the Clayton Antitrust Act, price discrimination for the most part is illegal.

Some forms of price discrimination still exist because they are seen as acceptable. You might have benefited from price discrimination the last time you went to a movie theater or flew on an airplane. Senior citizen, student, and military discounts are usually offered at theaters. Business travelers and vacationers often pay very different prices for tickets on the same flight even though they might both fly coach. How do the airlines get away with it? It is defensible because vacationers and business travelers have different elasticities of demand for airline tickets. Vacationers have elastic demand for tickets because they are able to book travel months in advance and are often willing to purchase nonrefundable fares. Business travelers' demand is much more inelastic, and thus they are willing to pay the higher price for the convenience of refundable fares and the privilege of booking tickets at short notice. By allowing airlines to charge different prices for basically the same ticket, the airline is able to better ration tickets between those who need a ticket and those who want a ticket.

The Good, the Bad, and the Ugly

Monopoly is not always a bad thing. Then again, it is not always a good thing. There are even times where monopoly is downright ugly, unless, of course, you are the proud parent of a monopoly. When monopoly is allowed to exist, it is for a good reason. Yet, even though the reason is good, monopolies can have some negative effects. History has shown that left unchecked, monopolies can harm an economy.

The Good

Good monopolies come in several forms. The first is natural monopoly. When the average cost of production falls as a factory grows larger, then economies of scale are present. **Natural monopoly** exists when economies of scale encourage production by a single producer. A commonly cited example of natural monopoly is your local electrical utility. A feature of power plants that encourages natural monopoly is that as the size of a power plant increases, the cost per kilowatt hour of electricity falls. You might own a small electrical generator. Imagine the cost of operating the generator or multiple small generators just to meet your home's electrical needs. Now imagine the cost of every household in a city running on multiple portable generators. The total fixed cost of generators for the community would be quite high and the variable cost of running gas or diesel generators would be astronomical.

Compare that situation with the one that most likely exists in your city. Instead of multitudes of portable generators, a few large coal-fired power plants are able to generate electricity for the entire city at a much lower total cost. Remember that utilities are monopolies. What keeps the utility from charging a monopoly price for electricity? Government regulates the prices that utilities are able to charge their customers for electricity. By controlling prices, government encourages low-cost production while allowing the utility to experience an accounting profit on production.

FACT

Besides patents, United States law also grants copyrights and trademarks to protect intellectual property. Copyrights protect written and other creative work and remain in force for seventy years from the original writing. Trademarks protect company and product names and can last indefinitely.

The technological monopoly is another form of monopoly that is encouraged. When a firm invents a new product or process and receives patent protection, the firm becomes a technological monopolist for that particular product. According to the United States Patent and Trademark Office, patent protection lasts for twenty years from the date on which it was originally applied. During that period of time, no other firm may develop or import the technology. The patent holder may develop or sell the rights to develop the technology to a firm that can legally operate as a monopolist.

During the period of patent protection, the patent holder as a monopolist can charge a monopoly price and earn economic profits. If monopoly prices are higher than competitive prices, why is this encouraged? Patent protection encourages innovation, invention, and research and development. Without the protection, firms would have little incentive to invest billions of dollars in research knowing the firm next door could just copy the product without having made the investment and profit nonetheless. It is because of patent protection and the ability to earn monopoly profits that American and European pharmaceutical companies develop so many lifesaving medications. Without patent protection, there would be little incentive for the pharmaceutical industry to pursue its research.

The Bad?

At times government may decide to step into the marketplace in order to provide a good or service. An example of government monopoly is the United States Postal Service. What about UPS and FedEx? Only the United States Postal Service is allowed to deliver a "letter" written on paper and delivered in a paper envelope. UPS and FedEx are in the package delivery business, even if that package is a letter written on paper and delivered in a flat paper-cardboard envelope.

Other examples of government monopoly include the various departments and agencies of the executive branch. Much of what they do and provide could be done by the private sector of the economy, but for many reasons the government has deemed them to be government functions.

The arguments for and against government monopolies fall mainly on philosophical grounds. Many conservatives and libertarians are opposed to government performing the functions of private enterprise on the grounds that government is wasteful and inefficient. Those with more populist viewpoints tend to see a need for government performing some of the functions of private enterprise on the grounds that government is less wasteful and more efficient. You decide.

The Ugly

Pure, unregulated monopoly is ugly. A firm that is the sole provider of a good or service is able to prevent competition. It can charge whatever price the consumer will pay. This is the monopoly that is most harmful to society. Although one may say "to the victor go the spoils," once-competitive firms that become monopolists need to be checked by regulation or broken up into competing firms. Competition benefits society by providing a variety of goods and services at competitive prices that accurately reflect the costs of production. **Pure monopoly** is the opposite of this condition. Pure monopoly is one good at a price that in no way reflects the true cost of production. The diamond monopoly of De Beers is the classic example of monopoly gone bad. De Beers at the height of its power dictated the diamond industry. By controlling the resource and coercing the wholesalers and cutters to abide by its demands, De Beers created an illusion of scarcity and value in the diamond market that allowed it to earn economic profits for over 100 years. Now, before you sell your diamonds in disgust, you should remember that you were not coerced to buy the diamond. You bought the diamond because the benefit outweighed the cost. The problem for the buyer is that you never realized how much of the cost was De Beers' profit.

CHAPTER 9: Government in the Marketplace

From time to time, people will petition government to step in and correct perceived wrongs in the market. Often this leads to unexpected results. Without considering how people might respond to incentives, well-intentioned policies can go astray. Because people are not going to stop acting like people, governments must consider whether or not their actions create perverse incentives. Likewise, there are times when the market fails to provide goods or properly assign costs. This calls for government intervention to either provide or redirect incentives in such a way that the market functions better.

Price Ceilings

In the early 1970s, America was faced with ever-increasing food prices. As a result, people clamored for the government to step in and halt the increases. Instead of considering the source of the problem and doing something about that, government attempted to treat the symptoms. In an effort to alleviate the suffering of households, the Nixon administration enacted price controls. One such price control was an effective price ceiling on food. Retailers could not charge a price higher than the government-mandated ceiling.

FACT

Rent controls were popular in the Northeast during the 1970s. Landlords responded by converting residential rental properties into office space or condominiums. Some landlords abandoned their properties because they were unable to maintain them under a system of rent controls.

A **price ceiling** is a legal maximum price for a good, service, or resource. At the time, the theory was that if the government imposed a price ceiling on food, prices would stop going up and everyone would have the food they wanted at the price they wanted. Of course, this assumes that people do not behave like people. Remember that prices are the result of the equilibrium of supply and demand. Also remember that these two forces are completely shaped by human nature.

The law of demand, which governs consumer behavior, says that as prices fall, consumers have an incentive to buy more, and as prices rise, consumers have an incentive to buy less. The law of supply, which governs producer behavior, says that as prices rise, producers have an incentive to produce more, and as prices fall, producers have an incentive to produce less. What effect does a price ceiling have on the market for food? Look at the incentives. A price ceiling encourages consumers to purchase but discourages producers from producing. Assume that meat is currently selling for $5.00 per pound. Consumers feel that the price is too high, so they petition government for a price ceiling of $3.00 per pound. Representatives, senators, and presidents all like to get re-elected, so they cater to consumers and enact the price ceiling. The $3.00 price signals to consumers to purchase more, but it signals to producers to produce less. The result of the price ceiling is a shortage of meat at the price of $3.00 per pound. At that price, more meat is demanded than is supplied. Consumers got a price ceiling of $3.00, but many consumers did not get any meat.

ESSENTIAL

Price controls are inefficient for many reasons. One reason worth considering is that they increase the need for monitoring and enforcement. That means increased government bureaucracy, which does not come cheap. Increased government spending equals more taxes or more borrowing.

Eventually, America abandoned price controls, but it took a decade to get the underlying inflation under control. Even today you still hear of people demanding that government cap prices of various commodities. As late as 2007, people were asking for price limits on gasoline. People continue to behave like people, and they still want low prices. Government and consumers would be wise to learn from the mistakes of the past and realize that attempts to control the market result in unintended consequences.

Price Floors

Consumers are not the only ones who ignore the basics of supply and demand. Producers have at times called for price floors. A **price floor** is a legal minimum price for a good, service, or resource. Probably the best-known price floor is minimum wage. In the market for resources like labor, households supply and businesses demand. Politicians representing areas with large populations of unskilled labor are often pressured by voters to increase the minimum wage. It's believed that an increase in the minimum wage is justified because employers will pay the higher wage and maintain the same number of workers. However, that works only if you assume that people do not behave like people.

For example, suppose that the city council of a major city, under pressure from voters, raises the minimum wage from the federal minimum to a city minimum of $10 an hour. Further assume that the equilibrium wage in the inner-city area was already $8 an hour and that in the suburbs it was $11 an hour. What will happen in the inner city and what will happen in the suburbs? In the inner city more producers (workers) will be willing to supply their labor at the higher price, but fewer consumers (employers) will be willing to employ that labor at the higher price. As a result, a surplus of unskilled labor develops, better known as unemployment. In the suburbs, the increase in the minimum wage has no effect, as the equilibrium wage was already $11. In the end, the policy meant to help the poor helped those who maintained their job but resulted in unemployment for those who were laid off and those who entered the job market looking for work at $10 an hour.

QUESTION

How important is minimum wage?

For all the attention it receives, it is not very important. In 2007, the Bureau of Labor Statistics estimated that only about 2.5% of wage earners made minimum wage.

Interestingly enough, those most in favor of increasing the minimum wage are often the same people who would be most harmed by the increase. Politicians know this now and will often pass increases in the minimum wage that keep it less than the average equilibrium wage for unskilled labor. For example, if the average market equilibrium unskilled labor wage is $8 an hour, then politicians will gladly increase the minimum wage from $6 to $7.50 knowing that it will have little economic effect. Yet, they can still put a feather in their cap for "raising" the minimum wage.

Taxes and Subsidies

Government has a power that businesses envy. It is the power to coerce payment, better known as **taxation**. The opposite of a tax is a subsidy. Government uses taxes and subsidies not only to raise revenue or redistribute income but also to shape peoples' incentives.

All effective governments have the power to tax. Taxes can be used as a tool of microeconomic policy. For example, if government wants to reduce the production of a certain good, they can tax the producer. This raises the cost of production for the producer and reduces the supply in the market. Prior to the Civil War, thousands of different forms of currency were in circulation as there was no national currency. The war provided impetus for a national currency. In order to ensure the success of the new greenback, congress taxed all the other forms of currency that had the effect of removing them from the market. How? Would you issue a dollar if it cost you $1.50?

Subsidies are used by government to encourage the production of certain goods and services. Many farmers and ranchers are subsidized by the government. Subsidies have the effect of increasing the supply of the good or service and reducing its price. Subsidizing farm goods ensures that there is always more than enough food and gives American farm exports a price advantage. Critics of farm subsidies argue that it creates inefficiency and misallocates scarce resources.

FACT

Many poor countries depend on agriculture as their chief export. A sticking point in WTO negotiations is that industrialized nations want poor nations to open up to free trade, yet industrialized nations are reluctant to end farm subsidies. These subsidies make it difficult for the poor nations to compete.

Subsidies can sometimes create a policy problem. They are much easier to award than they are to take away. Once a group benefits from a subsidy, they have a strong incentive to ensure that the subsidy continues. Political scientists refer to the relationship that forms as an iron triangle. Subsidy recipients support the election of members of congress on the relevant committees, and the committee members have oversight of the relevant bureaucratic agency that administers the subsidy.

Market Failures

There are times when the market fails to provide a necessary good or service, or fails to properly assign costs. Economists refer to this as a **market failure**. Public goods, positive externalities, and negative externalities are three forms of market failure. Government has the capacity to step in and deal with these market failures in a variety of ways.

Public Goods

There are times when the market does not provide a good or service that people want. If a good is **nonrival** and **nonexcludable**, the free market will probably not provide it. A good or service is nonrival when one person's consumption of the good or service does not diminish another's consumption of the good or service. For example, when you go to the movie theater, the presence of another person does not diminish your ability to consume the service, unless of course he has a screaming infant in his lap. A candy bar is an example of a **rival good**. If you eat the candy bar, then another consumer cannot. A good is excludable if the producer can withhold it from those unwilling to pay for it. Public highways are an example of a nonrival, nonexcludable good. A private firm has little incentive to produce a public highway at its own cost. Therefore, it is up to government to provide these types of goods.

Public goods and services include such things as roads, bridges, police protection, fire protection, national defense, and public education. Although private enterprise benefits from the presence of these public goods, individual firms have little incentive to provide them. Wal-Mart may benefit from the presence of Interstate Highway 35 as a means of transport, but Wal-Mart is not about to build a highway from Laredo, Texas, to Duluth, Minnesota. Why? IH35 is nonrival and nonexcludable. As such, Wal-Mart would find it next to impossible to profit by owning it.

Positive Externality

Sometimes the production of a good or service creates a spillover benefit for someone who is neither the producer nor the consumer, called a **positive externality**. Assume you live in a typical American neighborhood. If your neighbors were to landscape and remodel their home in such a way that significantly increased its value, you also would benefit. Your home would also appreciate in value, but you paid nothing for the increase. Flu vaccines create a positive externality as well. If you're concerned about catching the flu, you would go to your doctor and pay for a flu shot. Your decision to get a flu shot creates spillover benefits for the people around you. Your immunity reduces the chance that they will contract the disease even though they did not pay for it. Economists refer to these people as **freeriders**.

> **FACT**
>
> Some economists argue that subsidized loans and grants for college students are part of the problem, not the cure. The financial aid system, while well-intentioned, has the effect of increasing demand for college. As you know, increased demand leads to higher prices.

When production of a good or service creates a positive externality, there is never enough of it. In order to increase the desirable good or service, government might choose to subsidize its production. Government subsidizes public schools for this reason. Even though private schools exist, there are not enough private schools to educate the population. An educated population creates a positive externality, so government subsidizes education for all children. Businesses in America do not have to teach their workers to read the employee manual or compute math problems.

Negative Externality

Negative externalities occur when the production or consumption of a good or service creates spillover costs to society.

Pollution is an example of a negative externality of the production process. For example, a shoe manufacturer produces shoes, but it also produces air pollution that is released outside the factory. When the firm sells the shoes to its customers, the cost of the pollution is not factored into the price of the shoes. People living near the factory bear part of the cost of production in the form of pollution but do not receive payment for the shoes that the factory makes. When a firm's production creates a negative externality, there is too much of the good being produced. In a situation like this, government can tax the producer to reduce the amount of shoes, and pollution, produced.

Black Markets

Despite the best efforts of government, whenever and wherever government attempts to control prices or meddle with the forces of supply and demand, you can be sure a **black market** will develop. Markets develop when there is a supply of and a demand for a product. It does not matter what it is. If someone makes it and people want it, there is a market for it. Furthermore, if government intervention creates surpluses or shortages, some enterprising individual will step in and provide a means of circumventing government and clearing the market.

QUESTION

What is so black about black markets?

The term *black market* derives from the fact that illegal transactions were often carried out at night under cover of darkness. Even though many black markets occur in broad daylight, the term has stuck.

When government establishes a maximum price, shortages result for the good or service. This shortage is remedied by the market. In New York City, a system of rent controls dating back to World War II is still in place. As a result, very desirable apartments can be rented at extremely low prices. But at the low prices, very few apartments are available. People with significant income might be willing to rent some of these apartments but are unable to do so legally. The market has responded with black market subletting. A renter benefiting from the rent control will turn around and sublet the apartment to a renter willing to pay a much higher rent (which is still lower than average). Recall that voluntary trade is wealth creating, so both the new renter and the original renter benefit from the transaction. In addition, the landlord is no worse off than he was before the sublet occurred. Yet the city of New York actively pursues these illegal sublets. Every year, the city uses scarce resources to enforce this system of price controls. A solution would be to remove the rent controls and allow the market to determine the price of rent.

Sporting events also provide an example of the inefficient outcome created by price controls. Who wants to see an event more, someone willing to pay $50 or someone willing to pay $1,000? When ticket prices are limited, those willing to pay more are forced to compete with those who would actually receive less utility from seeing the event. Ticket scalpers perform an important function. By selling the tickets to the highest bidder, individual and societal utility is maximized. The original ticket seller received the price for which she sold the ticket. The person willing to pay the most ended up with the ticket voluntarily. The scalper received a profit for standing in line and turning an inefficient system into an efficient one. Yet, communities spend scarce resources trying to prevent ticket resale.

Markets also develop to thwart price floors. Workers willing to work for less than minimum wage might resort to contract labor or receiving illegal cash payments. Situations where workers are forced to accept low wages or face deportation are not mutually beneficial, nor are they voluntary. Exploitation occurs when one party to a transaction is unable to freely choose whether or not to engage in the transaction. Assuming that the transaction is voluntary, both business and workers benefit. If the transaction were not mutually beneficial, then it would not occur.

FACT

Increased use of credit cards, debit cards, and other forms of electronic payment have put a dent in black market activity. Most black market activity is dependent upon cash. As the use of cash diminishes, so does the anonymity that goes along with it. As a result, fewer black market transactions can occur.

Even food subsidies for needy families are subject to black market activity. Some receiving food assistance will willingly trade $2 in food assistance for $1 cash. They benefit because they now have the freedom to purchase what they want. The buyer benefits by purchasing groceries at half price. The problem with the system is that it is inefficient and creates disutility for the recipients. The taxpayer wins under a cash payment system because he is able to give the same value of service at a fraction of the cost. The recipient wins because he is able to buy groceries without the stigma of having to pay using food stamps, and if he so chooses he can buy other things he values more.

The obvious argument against a system like this is that some might not buy food with the food assistance but instead purchase alcohol, cigarettes, or even illegal drugs. Consider this example: assume that a person receiving food assistance also has an addiction to cigarettes. She receives $100 in food stamps and immediately sells them on the black market so that she can buy $50 in cigarettes. In the end, she has $50 in cigarettes and no food. Now assume that a person receiving food assistance receives $100 cash. He purchases $50 in cigarettes but now has $50 left for food. He is better off and the taxpayer is better off. In addition, the cash payment system removes the black market and is much less expensive to administer.

CHAPTER 10: Financial Markets

Have you ever wondered what the people on the news are talking about when they say something like, "The Dow Jones Industrial Average closed up 30 points today on heavy trading. The S&P 500 also edged higher. The NASDAQ was mixed. Foreign markets opened lower on news that the Fed will maintain near zero interest rates for the foreseeable future. Corporate bond prices sank as many issues were downgraded while the yield on the 10-year Treasury ended lower"? If you subscribe to the *Wall Street Journal* or *Financial Times*, or regularly watch CNBC, Bloomberg, or Fox Business Channel, you probably understand the lingo. But if you are like many Americans, the financial markets are a complete mystery. Even though they appear complicated, financial markets serve a very basic purpose: to connect the people who have money with the people who want money.

Loanable Funds Theory

Economists offer a simple model for understanding financial markets and how the real interest rate is determined. This hypothetical market, referred to as the **loanable funds market**, exists to bring together "savers" and "borrowers." Savers supply, and borrowers demand the part of savers' incomes that are not spent on goods and services. The real interest rate occurs at the point where the amount saved equals the amount borrowed.

According to the law of supply, producers are only willing to offer more if they can collect a higher price because they face ever-increasing costs. In the loanable funds market, the price is the real interest rate. Savers, the producers of loanable funds, respond to the price by offering more funds as the rate increases and less as the rate decreases. Borrowers act as consumers of loanable funds — their behavior is explained by the law of demand. When the interest rate is high, they are less willing and able to borrow, and when interest rates are low, they are more willing and able to borrow.

ALERT

Newcomers to economics are often confused by the use of the term **investment**. In economics, investment means borrowing in order to purchase physical capital. If the topic is stocks and bonds, then investment is understood to mean pretty much the same thing as saving. Savers engage in financial investment, which provides the funds for borrowers to engage in capital investment.

According to the expanded view of the loanable funds theory, savers are represented by households, businesses, governments, and the foreign sector. Borrowers also are represented by these same sectors. Changes in the saving and borrowing behavior of the various sectors of the economy result in change in the real interest rate and change in the quantity of loanable funds exchanged. For example, a decision by foreign savers to save more in the United States results in a lower real interest rate and a greater quantity of loanable funds exchanged for the country. A decision by the U.S. government to borrow money and engage in deficit spending would increase the demand for loanable funds and result in a higher real interest rate and a greater quantity of loanable funds exchanged. The loanable funds theory of interest rate determination is useful for understanding changes in long-term interest rates.

Liquidity Preference

John Maynard Keynes's **liquidity preference theory** explains short-term nominal interest rates. Instead of looking at saving and borrowing behavior as the determinant of interest rates, Keynes taught that short-term interest rates are a function of consumers' liquidity preference or inclination for holding cash. In Keynes's theory, the money market, as opposed to the loanable funds market, was central to explaining interest rates.

The money market is where the central bank supplies money, and households, businesses, and government demand money at various nominal interest rates. Central banks like America's Federal Reserve act as regulated monopolies and issue money independent of the interest rate. Liquidity preference is the demand for money. At high nominal interest rates, people would rather hold interest-bearing noncash assets like bonds, but as interest rates fall, people are more willing to hold cash as an asset because they are not sacrificing much interest to do so.

QUESTION

Why are there two theories of interest rate determination?

Economists have competing theories for many economic phenomena. Interest rates are just one of many concepts on which economists have differing points of view. The loanable funds theory is associated with classical economics, whereas the money market theory is associated with Keynesian economics.

If the Fed wants to lower the nominal interest rate to encourage investment and consumption, they increase the money supply, and if they wish to raise nominal interest rates in order to curtail investment and consumption, they decrease the available supply of money. Increases or decreases in the nominal gross domestic product cause the demand for money to either increase or decrease. Assuming that the Fed holds the supply of money constant, increases in money demand result in a higher nominal interest rate, whereas decreases in money demand reduce the nominal interest rate.

The Money Market

Firms, banks, and governments are able to obtain short-term financing in the money market. Securities with maturities of less than one year are included in this market. Businesses with excellent credit can easily borrow in the money market by issuing commercial paper, which is nothing more than a short-term IOU. For example, Big Corporation is expecting payment from its customers at the end of the month, but has to pay its employees before then. In this case, Big Corporation can issue commercial paper in exchange for cash with which to pay its employees. As soon as the customers make their payment to Big Corporation, the company can turn around and repay who-ever holds its commercial paper. For the company, commercial paper allows them to manage their cash flow, and for the lender it provides a liquid and relatively safe way to earn some interest on their extra cash.

FACT

You may have money saved in a money market mutual fund. In general, money market investments are extremely safe and liquid, which makes them an important part of most savers' portfolios. The fund's net asset value or price is one dollar per share and is never expected to change. On extremely rare occasions the net asset value may decline in what is called "breaking the buck." When this happens, savers lose, and it can precipitate a panic.

The United States Government auctions **treasury bills** (T-bills) for its short-term cash needs. The T-bills have various maturities of less than one year. Investors like T-bills because they allow them to earn risk-free interest while maintaining liquidity in case they need their cash quickly for other purposes. The government benefits because T-bills provide the government with easy access to cash for government spending.

T-bills are different from other forms of government securities in that they are sold at a discount from face value. T-bills have a face value of $1,000, but buyers pay less than that. The difference between the face value and the amount paid represents an interest payment. For example, if Rich Guy buys a 52-week $1,000 T-bill for $950, he will receive interest of $50 at maturity, which is equivalent to earning 5.26% on a $950 investment.

Banks meet their short-term financing needs by lending and borrowing federal funds, also known as fed funds. Banks may borrow from each other in the fed funds market to satisfy their legal reserve requirements or to meet their contractual clearing balances.

This component of the money market is important for maintaining bank liquidity. It also increases efficiency by encouraging banks to put all of their available excess reserves to work earning a return. Because banks keep most of their reserves on deposit with the Fed, the exchange of these federal funds occurs almost immediately as the banks exchange their reserve balances between each other. The Federal Reserve affects this market by influencing the fed funds rate, which is the interest rate banks charge each other for the use of overnight federal funds.

Bond Market

For long-term financing, governments and firms are able to borrow in the bond market. When investors buy bonds, they are lending money to sellers with the expectation that they will be repaid their principal plus interest. For bond issuers, the bond market provides an efficient means of borrowing large sums of money. For the buyer, bonds provide a relatively secure financial investment that provides interest income.

ESSENTIAL

> Markets function well when secondary markets exist. The stock and bond markets that you hear about on the news are secondary markets. In secondary markets, buyers and sellers are trading in used goods and services. People are much more likely to buy stocks when they are initially offered for sale if they know that they can sell them later in the secondary market. Markets would not function well if consumers were unable to turn around and profit from their financial investments.

Types of Bonds

You are probably familiar with two types of bonds. **Coupon bonds** are sold at or near face value and provide guaranteed interest payments. **Zero-coupon bonds** are sold at a discount from face value and pay face value at maturity. Either type makes an attractive investment for people seeking interest income while preserving their principal. The ability to sell bonds on the secondary market makes them relatively liquid, which is also important to investors.

The United States government issues several types of bonds with maturities greater than a year. Treasury notes and treasury bonds are a primary source for financing the federal budget. **Treasury notes** are medium-term securities with maturities ranging from two to ten years. **Treasury bonds** are long-term securities that mature after thirty years. The interest rate on the ten-year Treasury note is important because it serves as a benchmark interest rate for both corporate bonds and mortgages. As the interest rate on the ten-year Treasury note fluctuates, corporate rates and mortgage rates fluctuate as well.

In addition to the treasury, independent agencies of the United States government are able to borrow by issuing bonds. Although they lack the guarantee of repayment that treasury securities have, agency securities are backed by the government and as such are seen as virtually guaranteed. The Federal National Mortgage Association (Fannie Mae), the Federal Home Loan Mortgage Corporation (Freddie Mac), and the Student Loan Marketing Association (Sallie Mae) are well-known agencies that issue bonds in order to finance their operations. Agency securities provide an alternative for investors looking for the security of government bonds but with higher interest rates.

State and local governments are also able to borrow through the bond market. **Municipal bonds** often finance schools, roads, and other public projects. The interest paid on the municipal bonds is exempt from federal income taxes, which makes them attractive to investors. Because the interest is tax exempt, municipal bonds do not have to offer as high an interest rate to attract investors. As a result, state and local governments are able to borrow more cheaply than the private sector.

Firms are able to borrow in the bond market by issuing corporate bonds. **Corporate bonds** provide businesses with the money they need for capital investment without having to arrange bank financing. In addition, corporate bonds allow for businesses to obtain funds without diluting ownership in the company. The chief advantage of bonds is that they provide firms with financial leverage. For example, if a company has $1,000 to invest in capital and can expect a return of 10%, the company will earn $100 from the investment. If, however, the firm borrows $1,000,000 and invests in capital that returns 10% a year, the firm is able to earn $100,000 without risking its own money. Because firms lack the ability to tax to repay bonds, investors require a higher interest rate on corporate bonds than on treasury and municipal bonds to offset the increased risk of default.

FACT

A relatively new type of bond, called a **collateralized debt obligation** (CDO), is seen as one of the culprits in the recent financial crisis. Traditionally, banks would lend money to people after carefully scrutinizing their ability to pay. Once approved, banks would then hold the customer's loan. Today, banks issue credit and then in a process called securitization, sell bundles of loans or CDOs to investors. This frees the bank to issue more credit.

Bond Risks

Bonds are not without their downsides. Investors face investment risk, inflation risk, interest rate risk, and the risk of early call. Since the possibility exists that governments or firms may fail to pay back their borrowed money, all bondholders face investment risk. If the rate of inflation increases during the life of a bond, the investor's return is offset by the inflation. If interest rates increase during the life of a bond, the value of the bond decreases until its effective yield equals the new higher interest rate. For bondholders, this means that they might lose principal if they try to sell it before maturity. If interest rates decline during the life of a bond, the issuer may find it beneficial to retire or call the old bonds and refinance at the new lower interest rate. For bondholders, this means that they lose out on earning the higher interest they would have had when the bond matured.

Prospective investors rely on rating agencies to determine the quality of the bonds. Moody's, Standard & Poor's, and Fitch rate bonds from "prime" to "investment grade" to "junk," and all the way to "in default." As a bond's rating falls, the issuer must reward the investor with a higher interest rate to compensate for the additional risk. The rating agencies provide a useful service and provide much-needed information for the consumer. However, consumers should not rely only on bond ratings. Serious questions have arisen from the 2008 financial crisis about the rating process. During that time, many bonds were so quickly downgraded that investors who believed they were holding prime-rated bonds discovered that they had junk bonds within a short period of time.

Interest Rates Revisited

Interest rates are made up of several components: the real rate, expected inflation premium, default risk premium, liquidity premium, and maturity risk. The real rate and the expected inflation premium make up the risk-free rate of return. This risk-free rate of return acts as the benchmark on which all other interest rates are based. Today the various bonds issued by the United States Treasury are the proxies for the risk-free rate of return.

Treasury securities have this distinction because the United States has never defaulted on its debt in its 220+ year history, and the secondary market for U.S. treasury securities is considered "deep" because it is backed by the full faith and credit of the United States. The importance of the secondary market in the treasury cannot be understated. Because so many governments, banks, businesses, and individuals desire U.S. treasury securities as a risk-free place to park their money, a condition is created where there is no doubt to the liquidity of the securities. The only appreciable risk faced by the holders of treasury securities is maturity risk. The longer the term of a bond, the greater the chance that interest rates will change from the one that existed at the time of purchase. If interest rates were to unexpectedly rise during the life of the bond, the value of the bond would decrease. As new bond prices fall, the effective interest rate on the bonds increase, which makes previously issued bonds less attractive. This, in turn, makes them less valuable.

Stock Market

Of all the financial markets, none receives as much media coverage as the stock market. Unlike bond markets, where investors are making loans to governments and firms, the **stock market** is where investors are able to purchase partial ownership in firms represented by shares of **stock**. Firms are able to raise the money they need for capital investment by issuing stock in an **initial public offering** (IPO). Investors purchase the stock with the expectation that it will either pay dividends or earn capital gains. Investors earn **dividends** when a company divides a portion of the profits among all of the owners according to the number of shares each owns. For example, if there are 100 shares of stock for a company, and the firm earns profits of $1,000,000 and decides to distribute half of the profits to its shareholders while reinvesting the other half, each share will earn a dividend of $500,000/100, or $5,000. Stock earns **capital gains** when it is sold at a higher price than for what it was purchased.

QUESTION

What are stock options?

Stock options are a type of derivative that allow for the purchase of shares of stock at a predetermined price. Companies often issue stock options to key employees as a reward for performance. The recipients can either sell their option contract on the options exchange or wait and exercise their option when the share price of the stock increases.

The majority of stock purchases and sales occur in the secondary market. When you place an order to buy stock, you are most likely buying shares that were previously owned by another individual. If

Tina buys Coca-Cola stock in the market, she is buying it from someone else, not Coca-Cola. If she pays $150 for two shares of Coca-Cola stock, some other investor who sold the stock will receive $150, but Coca-Cola will receive nothing. The only time the firm receives money in a stock purchase is through an IPO or when the firm sells stock that it had repurchased earlier.

Companies issue two types of stock, common or preferred. **Common stock** provides investors with partial ownership of a firm and also grants them the right to vote for the firm's leadership. **Preferred stock** also provides an ownership claim on a firm but does not allow for voting privileges. Preferred stock is so named because preferred stockholders get paid before common stockholders when it comes time to pay dividends. The decision to issue common stock or preferred stock is influenced by the possible downsides of each. Because common stock allows shareholders a voice in corporate governance, the original founders of a firm might find themselves displaced if new common stockholders gain either a majority or plurality of the shares and elect new leadership. Because preferred stock guarantees dividends to shareholders, firms seeking to grow by reinvesting profits in the company might be hindered by this liability.

So What's the Point?

Ultimately the function of these various markets is to allow savers to connect with borrowers. Businesses seeking to expand their capital investment look to the bond market and the stock market as a source of needed funds. Firms are very conscious of their operating cost, so finding the appropriate combination of stock and bond financing, or capital structure, with which to finance investment is important both to the business and the economy as a whole. There is an inverse relationship between the interest rate and investment.

The lower the interest rate, the lower the cost of capital, and the more firms are able to invest in physical capital. Likewise, as the interest rate increases, the cost of capital increases, and firms become less willing to invest in physical capital.

Are low interest rates enough to encourage investment? Unfortunately, the answer to that question is no. A firm's decision to invest in capital is also conditioned on the expected rates of return. If firms expect higher rates of return, they are more willing to invest at each and every interest rate. If firms think there is little chance to earn a profit, then they are much less willing to invest in capital.

Here's a way of looking at the effects of interest rates and expected rates of return on a business. Imagine that the expected rate of return is the pressure applied to a car's gas pedal and the interest rate is the pressure applied to the brake pedal. Increased profit opportunities are represented by pressing down on the gas pedal and decreased profit opportunities are represented by letting your foot off of the gas pedal. Similarly, increased interest rates are represented by pressing down on the brake pedal, lower interest rates are represented by putting less pressure on the brake pedal, and zero interest rate is analogous to not putting any pressure on the brake pedal. If interest rates are low and expected profits are high, the business moves forward with investment and grows. If, however, the interest rate is higher than the expected rate of return, firms are stationary. In the end, increases in the expected rate of return will accelerate investment while increases in interest rate will slow investment or even bring it to a complete stop.

CHAPTER 11: Foreign Exchange and the Balance of Payments

Whether you know it or not, foreign exchange is a part of your everyday life. From the products you buy to the vacations you take, foreign exchange affects much of what you do. The flow of currency between nations is also a matter of record keeping. The balance of payments records all of the inflows and outflows of currency from a country. The sum of net exports, net foreign factor income, and net transfers is the current account balance, while net foreign investment and official reserves make up the financial account.

Exchange Rates

What determines the price of apples? If you answered supply and demand, then you would be correct. What determines the foreign exchange price of a dollar? Once again, if you said supply and demand, then you are right on target. Whenever one currency is exchanged for another, foreign exchange has occurred and an exchange rate has been paid. The **exchange rate** is nothing more than the current price of a currency in terms of another currency. For example, $1.00 may buy you €0.73, £0.64, ¥89.41, or SFr 1.07.

Exchange rates are determined in the world's largest market. Annual trade volume approaches $1 quadrillion (a thousand trillion), with transactions occurring twenty-four hours a day. The foreign exchange market is dominated by the British, Americans, and Japanese with the vast majority of trades occurring in the U.S. dollar. The euro (€), pound sterling (£), Japanese yen (¥), and Swiss franc (S_{Fr}) are the other hard currencies most often traded. Large banks are the main players in the foreign exchange market, brought together through a system of interconnected brokers. The banks serve both corporate and individual customers who need foreign exchange in order to conduct business. Central banks also participate in the foreign exchange market in order to either manipulate exchange rates or to correct imbalances between their county's current and financial accounts. Finally, speculators looking to profit from arbitrage opportunities also participate in this immense market.

ESSENTIAL

When traveling, the best exchange rates can often be found on your credit card or by accessing an ATM machine. Banks are the major players in the foreign exchange market and are able to grant their customers some of the most competitive rates.

Exchange rates are subject to the forces of supply and demand. As exchange rates rise, people are more willing to sell their currency, but others are less willing to buy, and vice versa. Appreciation occurs when an exchange rate increases, and depreciation occurs when exchange rates fall. Changes in exchange rates can wreak havoc on businesses and entire economies. Economists and policymakers must therefore consider the implications of their actions on exchange rates. Tastes, interest rates, inflation, relative income, and speculation affect exchange rates and thus economies as a whole.

Taste and Preference

As consumers' tastes for imported goods change, so does the exchange rate between the countries involved. The popularity of cars and consumer electronics from South Korea among American consumers necessarily creates a demand for South Korea's currency, the won (.). In order for American importers to acquire South Korean products, they must supply dollars and demand won in foreign exchange. The result of this change in tastes is a depreciation of the dollar and appreciation of the won.

Over time, appreciation of a country's currency may reduce the popularity of its products as they become relatively more expensive. Given that people often prefer relatively cheaper goods, foreign manufacturers that export goods to the United States might choose to price their products directly in dollars as opposed to their domestic currency. This helps to insulate the manufacturer from changing exchange rates, which might reduce their product's competitiveness on the American market. German automakers engage in this practice to offset the effects of the euro's relative strength against the dollar.

Interest Rates and Inflation

Changes in real interest rates also affect the foreign exchange market. Unlike trade flows, which tend to be rather stable, changing interest rates can cause sudden fluctuations in exchange rates. Savers are attracted to high interest rates, so when one country's real interest rate rises relative to another country, savings flow toward the higher interest rate. With all other things equal, increases in American interest rates relative to Japanese interest rates will cause an increased supply of yen and increased demand for dollars as Japanese savers seek to earn higher American interest rates.

The result is an appreciated dollar and depreciated yen. Eventually, the dollar's appreciation relative to the yen will offset any gains made off of the interest rate difference. This is referred to as **interest rate parity**.

The presence of inflation in an economy provides an incentive for its people to exchange their currency for one that is more stable. Zimbabwe's inflation has reduced the value of its currency to the point where a sheet of paper is more valuable than a single unit of its currency. Rational Zimbabweans trying to store value will eagerly supply the Zimbabwe dollar for any available currency, like South Africa's rand. The result of inflation is to not only reduce the value of the currency domestically but also in foreign exchange.

ALERT

Foreign exchange has a definite cost attached to it. If you choose to participate in the foreign exchange market, know that you will likely face different exchange rates in buying and selling currencies. So before you open up a CD in a French bank, consider the cost of purchasing and selling your euros before cashing in on a higher interest rate.

Relative Income and Speculation

Probably the most counterintuitive outcome results from differences in economic rates of growth. As an economy's income increases relative to another country, the exchange rate between the two changes. For example, if Canada's income increases relative to the

United States' income, the Canadian dollar weakens relative to the U.S. dollar. Why? As Canadians experience higher incomes, their propensity to import also increases. In other words, as you get richer, you go shopping more often and supply more currency in foreign exchange. The counterintuitive outcome is that as an economy strengthens, its currency weakens, and as an economy weakens, its currency strengthens.

Banks, companies, and individuals attempting to profit from foreign exchange will speculate in the market. For example, a speculator may suspect that European interest rates will rise faster than American interest rates, so she will purchase euros and hold them until they have appreciated enough to show a profit. For most people, however, speculating in foreign exchange is about as profitable as playing craps in Las Vegas.

Although speculating is risky, speculators do perform an important economic function by bearing exchange rate risk. Risk-averse companies engaging in foreign exchange would rather know exactly what exchange rate they will be paying or earning when conducting transactions. These companies will purchase a futures contract that guarantees a specified exchange rate in order to avoid sudden shifts in exchange rates that might reduce the profitability of the underlying transaction.

ESSENTIAL

> Over the course of a year you might hear experts lamenting the size of the trade deficit, while later others might complain of a weak dollar. Actually, a weak dollar is the remedy for a trade deficit. Depreciated dollars

encourage exports and discourage imports. Words like strong and weak can be misleading. They do not necessarily mean good or bad.

The Law of One Price and Purchasing Power Parity

Specialization and trade are based upon comparative advantage, differences in capital and labor, and diversity in consumption. Exchange rates, however, should not determine specialization. According to the economic **law of one price**, or **purchasing power parity**, after accounting for the exchange rate, the prices of similar goods should be the same regardless of where they are purchased.

FACT

The *Economist* magazine prints a yearly index of purchasing power parity called the Big Mac Index. The index allows you to see the cost of a McDonald's Big Mac in various countries adjusted for exchange rate differences. A Big Mac in Switzerland might cost you more than $4.00, while one in China will only cost $1.50.

Assume that the exchange rate between the U.S. dollar and Swiss franc is such that $1.00 = S_{Fr} 1.10. In the long run, a pair of snow skis that sell for an average price of $500 in the United States. should therefore sell for an average of S_{Fr} 550 in Switzerland. What if the skis were selling in Switzerland at an average price of S_{Fr}

600? Swiss consumers would benefit from importing the skis from America, supplying Swiss francs and demanding U.S. dollars in foreign exchange until the exchange rate reflected purchasing power parity.

Net Exports

In the United States, most producers focus on meeting the demands of the domestic market. Some, however, produce goods and services for export to foreign markets. Other businesses import those goods and services for which there is a demand or the United States does not produce. Half of the United States's trade occurs with Canada, Mexico, China, Japan, Germany, and the United Kingdom. **Net exports**, or the balance of trade, is equal to the value of all exports minus the value of all imports. Net exports in the United States are negative because the value of imports exceeds the value of exports. This is referred to as a **balance of trade deficit**.

Other countries, like China, Germany, and Japan, have balance of trade surpluses because the value of their exports exceeds that of their imports. Surprisingly, even though the United States has a balance of trade deficit, it is still the world's largest exporter with 12% of the global share, compared to China with only 6.4%.

The balance of trade can be further broken down into the balance of trade in goods and the balance of trade in services. For the United States, the balance of trade in goods is what contributes to the trade deficit. Americans have a preference for foreign consumer goods and resources. On the other hand, the United States tends to be a net exporter of services. America's transportation services and logistical know-how, as well as engineering, legal, and other technical services, are exported to the rest of the world.

The Commerce Department's Bureau of Economic Analysis (BEA) is the government agency responsible for measuring the balance of trade. According to the BEA, in 2008 the trade deficit measured approximately $696 billion. This total trade deficit was composed of an $840 billion trade in goods deficit combined with a $144 billion trade in services surplus.

Net Foreign Factor Income

Whenever land, labor, capital, and entrepreneurship are employed, they receive payments of rent, wages, interest, and profits. It does not matter whether the factors are employed domestically or abroad. When factors of production are employed abroad or the owners of those factors reside abroad, then the factor payments require foreign exchange. The use of foreign capital requires that interest payments be submitted to the country of origin. This represents a **foreign factor payment**. When domestic factors of production are employed abroad, they earn **foreign factor income**. Wages paid to American workers abroad from foreign employers are an example of foreign factor income. In addition, dividends earned by Americans holding shares of stock in foreign companies are included as foreign factor income. Regardless of whether it is an inflow of foreign factor income or an outflow of foreign factor payments, foreign exchange is required.

Net Transfers

Other actions besides trade require foreign exchange. Transfer payments, called **remittances**, and foreign aid create outflows of currency from rich countries to poor countries in foreign exchange markets. India is a prime example of a country dependent upon a large expatriate community. Many Indians living abroad remit money to their extended families back in India. Likewise, Mexico is heavily dependent upon remittances from emigrants working in the United States. As a matter of fact, remittances are the second-largest source of income for Mexico after oil exports. As of 2006, Mexico and India received remittances of $25 billion each, much of which originated from the United States. Remittances differ from foreign factor income in that they are unearned by the recipients. Foreign aid in the form of cash payments also creates a need for foreign exchange. The United States's aid payments to Pakistan, Israel, and Egypt are examples.

Net Foreign Investment

Unlike net exports, net foreign factor income, and net transfers, which involve one-time exchanges, **net foreign investment** creates recurring payments and income. When citizens of one country purchase the real or financial assets of another country, it is classified as net foreign investment. The goal of foreign investment is to earn foreign factor income in the form of interest and profits. A decision by Italian entrepreneurs to buy land and develop a theme park in Iceland, although not the best idea, would create an inflow of euros to Iceland with the purpose of creating an outflow of kronur to Italy as foreign factor income.

Net foreign investment also includes portfolio investment. **Portfolio investment** occurs when foreigners purchase the financial assets of another country. During the recent financial crisis, the dollar appreciated quickly as foreign investors sought the safety of U.S. treasury bills, notes, and bonds. Purchasing shares of stock in foreign companies is another form of portfolio investment. Because developing countries often have higher rates of economic growth and higher rates of interest, financial investors might purchase bonds and stocks in these countries.

QUESTION

Which is better, a weak dollar or a strong dollar?

It depends. Are you importing or exporting? Importers benefit from a strong dollar because it makes foreign goods relatively cheap. Exporters benefit from a weak dollar because it makes U.S. goods relatively cheap. So if you are going on a foreign vacation, hope for a strong dollar while you are there, and then hope it weakens on the flight back so that you profit on the exchange when you return.

Foreign investment is an important source of savings for the host country. These savings can be used to invest in physical capital and expand a country's economy. They can also be used to finance a trade deficit, as is the case with the United States. Currently, the United States is able to enjoy cheap Chinese imports in excess of United States exports to China because of a counterbalancing flow of savings from China to the United States. For every dollar spent on Chinese imports, there is a return flow as the Chinese use the same dollars to purchase both American exports and also American financial assets. This brings up an important point. The balance on the current account is completely offset by the balance on the financial account.

Foreign investment is not without its downside. Just as easily as savings can flow into a country through portfolio investment, the savings can flow out. A sudden outflow of savings may precipitate a currency crisis, inflation, and high interest rates for the affected country. This phenomenon has occurred before. The 2001 Argentine financial crisis is a notable example. Argentina defaulted on its foreign debt, and soon savings flowed out, which led to financial troubles and political turmoil as well.

Countries that fear capital outflows often enact capital controls. **Capital controls** limit the ability of savings to flow out of a country. During the Asian financial crisis of the late 1990s, India fared rather well as capital controls limited the outflow of savings from the country. The downside of capital controls is they provide a reason for foreign investors to avoid placing their savings in a country. The mere mention of capital controls is sometimes all it takes for capital outflow to occur.

Official Reserves

In addition to foreign investment, the financial account of the balance of payments includes **official reserves**. Central banks, like the United States Federal Reserve System, maintain reserves of foreign currency or official reserves. The purpose of official reserves is to provide a stabilizing influence in the foreign exchange market. If a balance of payments deficit occurs, the Federal Reserve reduces its foreign reserves in order to zero out the balance. In the case of a balance of payments surplus, the Federal Reserve acquires additional foreign reserves to zero out the balance.

The United States is in the unique position of issuing the reserve currency for the majority of countries as most countries' foreign reserve holdings are in U.S. dollars. Lately, the size of China's and Japan's official reserves has been a cause of concern for many in the financial community. As of 2010, these Asian countries have over $3 trillion worth of reserves. Some politicians, economists, and financial experts fear that if China or Japan were to reduce their dollar holdings, a collapse could ensue from a sudden increase in the supply of dollars in foreign exchange. Others contend that this would be every bit as harmful to the Chinese or Japanese, and thus they have little incentive to dump their dollars.

To Fix or Float? That Is the Question

At the end of World War II, countries wanting to promote international cooperation and reduce the economic incentives for war met in Bretton Woods, New Hampshire, and established a **fixed exchange rate** system pegged to the dollar. The benefit of such a system was that businesses could easily engage in foreign trade without fear of losing money from fluctuating exchange rates. In addition, by pegging currencies to the stable dollar, foreign governments were responsible for practicing sound economic policies such as not creating inflation by recklessly printing currency. The Bretton Woods system was at first very successful, but by 1971 it had completely collapsed. Today many approaches to exchange rates exist. Some countries let their currency float, others peg their currency to the dollar, and still others have unified under a single currency.

The United States and United Kingdom allow their respective currencies to float on the foreign exchange market, which means that they do not use official reserves to maintain exchange rates. The benefit of this system is that it gives policymakers the ability to practice interest rate policies to encourage domestic growth or slow inflation without having to consider exchange rates.

FACT

China's exchange rate policy is a political hot potato. On one hand, American consumers benefit from the low prices on imported goods. On the other hand, domestic producers find it hard to compete against China's prices.

China, Japan, and Hong Kong, however, peg their currencies to the U.S. dollar by actively trading their holdings of foreign reserves. This allows their respective countries to maintain a competitive advantage against others in the American market. China's exchange rate policy is what keeps Chinese exports relatively cheap. Under a **floating exchange rate**, American demand for Chinese goods would force up the exchange rate and eventually make Chinese goods less desirable. The downside for China is that all of the money spent on stabilizing the exchange rate could probably be put to more profitable use. Also, managing the exchange rate means acquiring more and more dollar-denominated financial assets, which makes China vulnerable to America's economic troubles.

The turn of this century saw another approach to exchange rate policy. Europe unified its economy under a single currency, the euro. France and Germany, for example, can now trade freely without regard to exchange rates. The advent of the euro created the world's second-largest currency after the dollar. For all of its benefits, the euro suffers a major drawback. Although Europe is united in name, the countries are quite distinct and have different economic conditions. The presence of a single currency managed beyond the national level means that the nations of Europe are unable to practice independent monetary policy to address economic problems unique to their specific countries. In 2010, Greece suffered a severe financial blow when its sovereign debt's credit rating was reduced. Unable to implement independent monetary policy, the Greeks have been forced to reduce their spending and raise taxes or risk default. The EU's reaction to this crisis will either weaken or bolster the euro's chances of becoming the dominant world currency.

CHAPTER 12: The Circular Flow of Economic Activity

In middle school science classes across the country, students are introduced to the idea of systems and how matter flows through them. The water cycle is an example of one of these systems. Water evaporates from the land and the oceans, condenses in clouds, precipitates over the land, percolates through the ground, or runs off in streams and rivers back to the oceans where it again evaporates. What is the source of energy for the water cycle? The sun. The economy can be understood in a similar manner. In an economic system, energy is derived from people's self-interest manifested in markets and through the political process. The circular flow model shows the relationship between households, firms, governments, and the foreign sector as they all interact in product, factor, and financial markets.

The Private Sector

The starting point for understanding the entire economy is a very simple model that illustrates how households and firms interact in the product market and the factor market. The **private sector** is nothing more than the households and businesses in an economy. Households buy goods and services from firms in the product market. Where did the households get the money to buy the goods and services? Households sell their labor and entrepreneurship to the firms in exchange for wages and profits. Households also sell natural resources to the firms in exchange for rents. If you work for someone else, you are selling your labor in the factor market so that you can get the goods and services you want and need in the product market. As a business owner or entrepreneur, you sell your entrepreneurial ability in exchange for profits from the firm. Firms employ land, labor, capital, and entrepreneurship in order to provide goods and services.

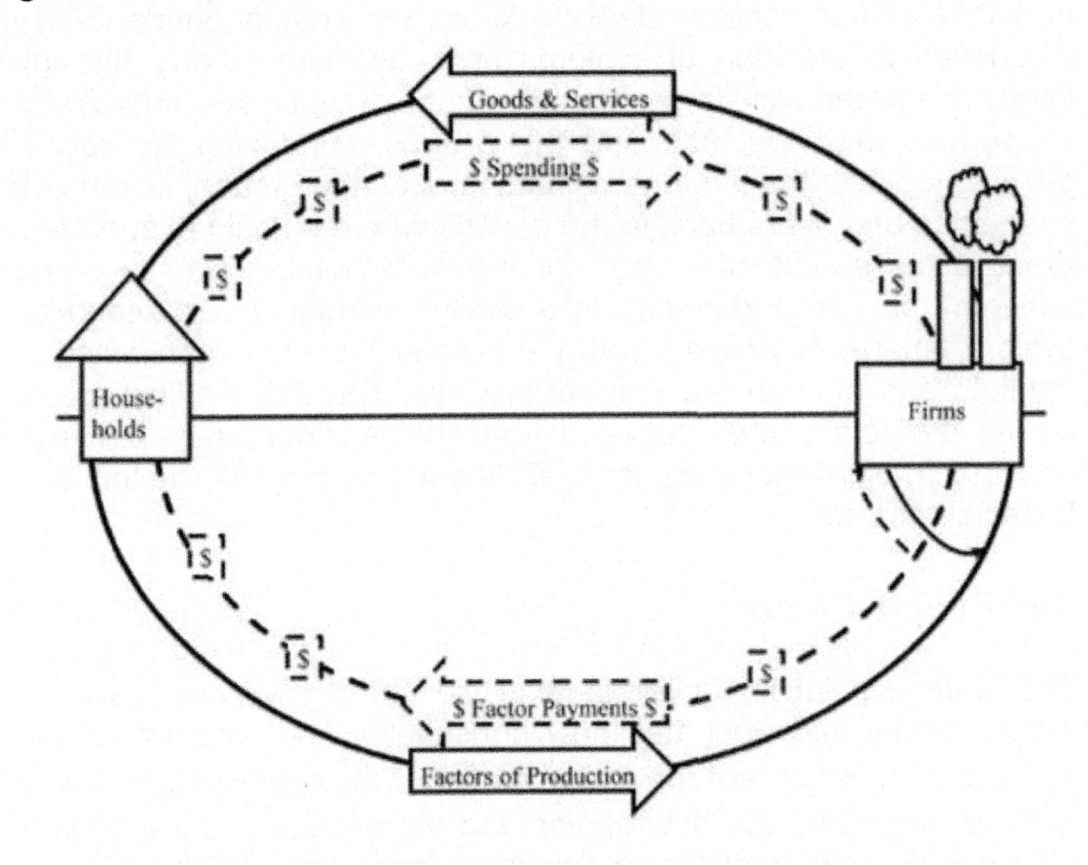

An economy made up of just the private sector. The small arrow leaving the factory in the factor market represents the flow of capital, and the arrow returning to the flow of factor payments represents interest payments occurring between firms.

There are two sets of flows between households and firms. Flowing in one direction are goods and services in the product market, and land, labor, capital, and entrepreneurship in the factor market. Consumption spending flows in the opposite direction of the goods and services. Flowing opposite the factors of production are factor payments, which include rent, wages, interest, and profits.

The Public Sector

Notice that the simple model of the private sector reveals a gaping hole in the middle. That void is filled by the public sector. The **public sector** refers to all levels of government, from local to federal. The public sector interacts with households by purchasing some of the factors of production in exchange for the factor payments. Government also interacts with firms by buying goods and services in the product market. Government combines the factors of production with the goods and services it buys from firms in order to provide public goods and services to the private sector. National defense, police, fire protection, schools, libraries, and roads are examples of the types of public goods and services provided by the public sector.

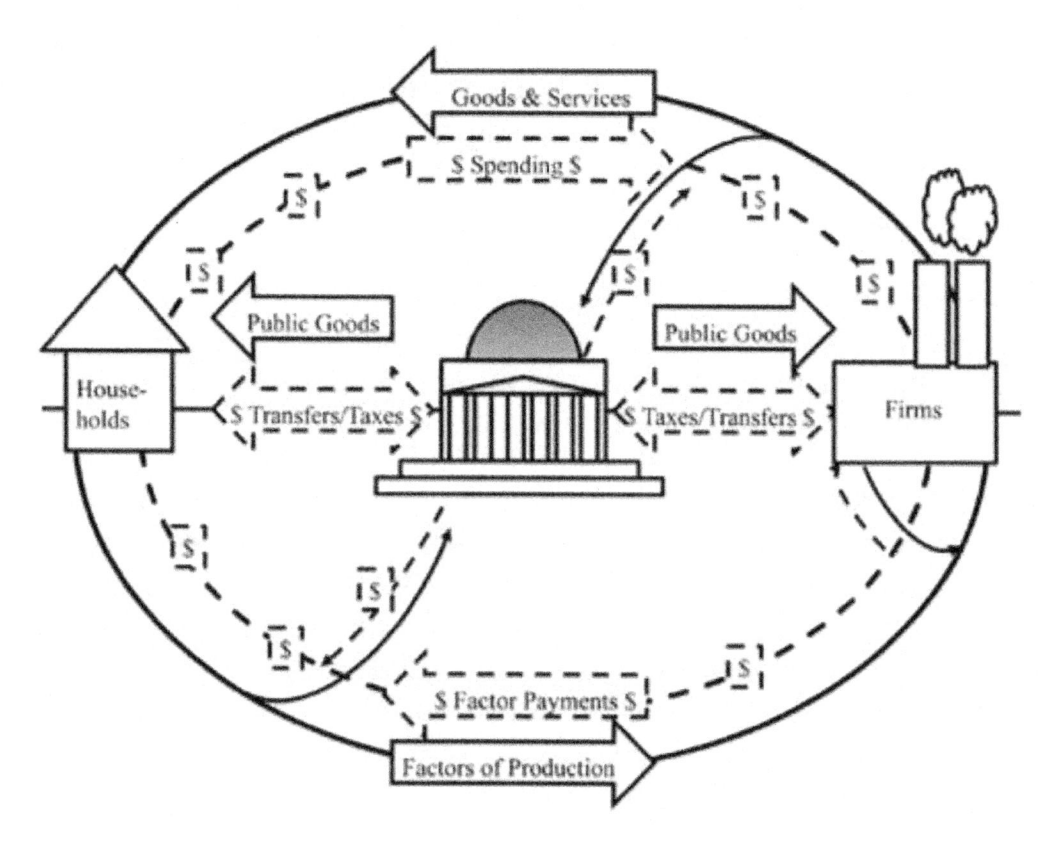

An economy with a private and public sector, or closed economy.

Where is the government getting the income with which to buy the factors of production from households and goods from the firms? Taxes, in both the factor market and the product market, are the source of governments' income. Taxes on personal income and corporate profits are collected in the factor market. Sales tax is an example of a tax collected in the product market. Sometimes the government subsidizes firms, which represents a flow of money from government to firms. Like firms that receive subsidies, many households receive transfer payments, like Social Security and welfare.

The Foreign Sector

Of course, the economy does not exist in isolation. The **foreign sector** refers to the rest of the world. The United States is part of the much larger world economy. The U.S. economy interacts with the rest of the world in both the product and factor markets. In the product market, not all goods and services flow to government and households. Some of the domestically produced goods and services are exported to people in other countries. In the same manner, not all of the goods and services that are purchased by households and government are domestically produced. Americans import goods and services from the rest of the world.

Not only do Americans trade goods and services with the rest of the world, they also trade the factors of production with the rest of the world. Land, labor, capital, and entrepreneurship flow from the rest of the world in exchange for foreign factor payments. Also, the factors of production flow to the rest of the world in exchange for foreign factor income. If a U.S. citizen earns income abroad, that is foreign factor income. Payments to foreigners for the use of the factors of production incur a foreign factor payment.

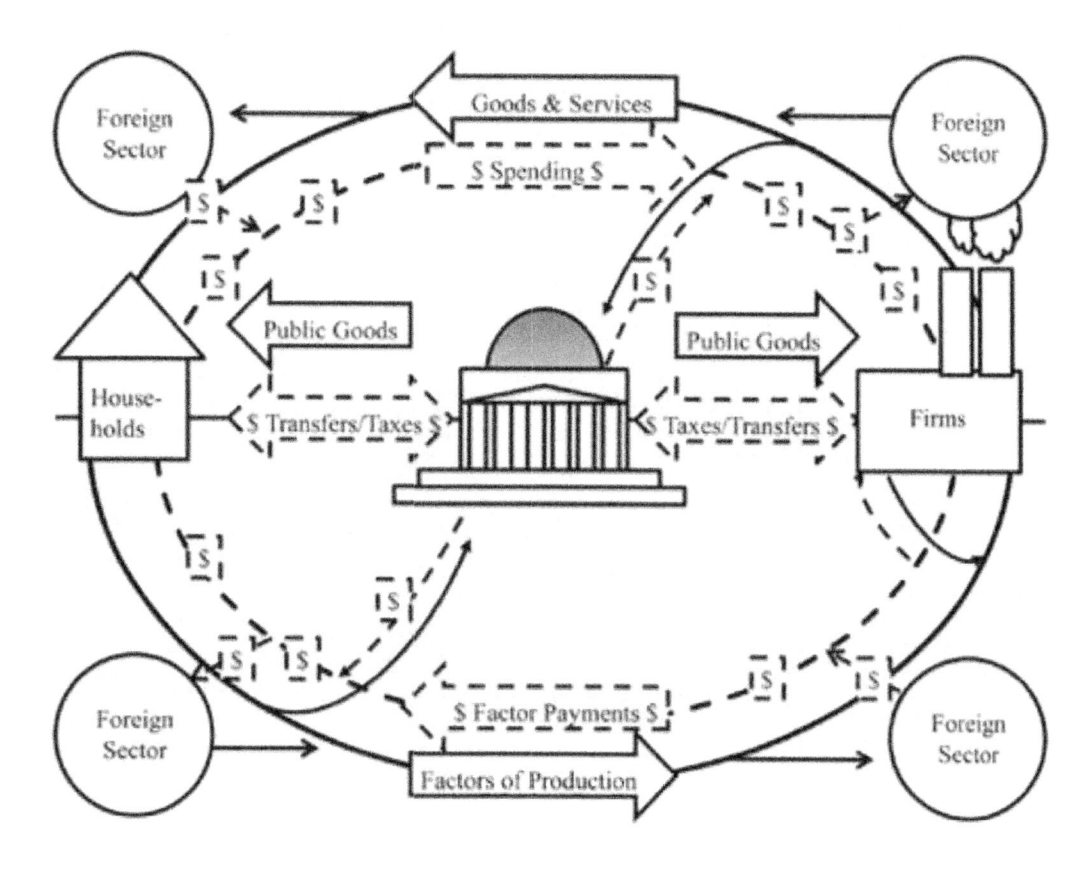

An open economy.

The Financial Sector

So far, financial intermediaries have been left out of the picture, literally. The **financial sector**, or intermediaries like banks, credit unions and stock exchanges, help to facilitate all of the transactions that have been mentioned in this chapter. The importance of financial intermediaries cannot be underestimated. They stand in the middle of almost all transactions.

Most of the transactions that are conducted in the product market either involve the use of checks, debit cards, credit cards, cash drawn from ATMs, or in the case of exports and imports, foreign exchange. All of these services are provided by the financial intermediaries. In the factor market, the story is the same. Most working Americans receive a paycheck, not cash. This means that, once again, financial intermediaries are involved.

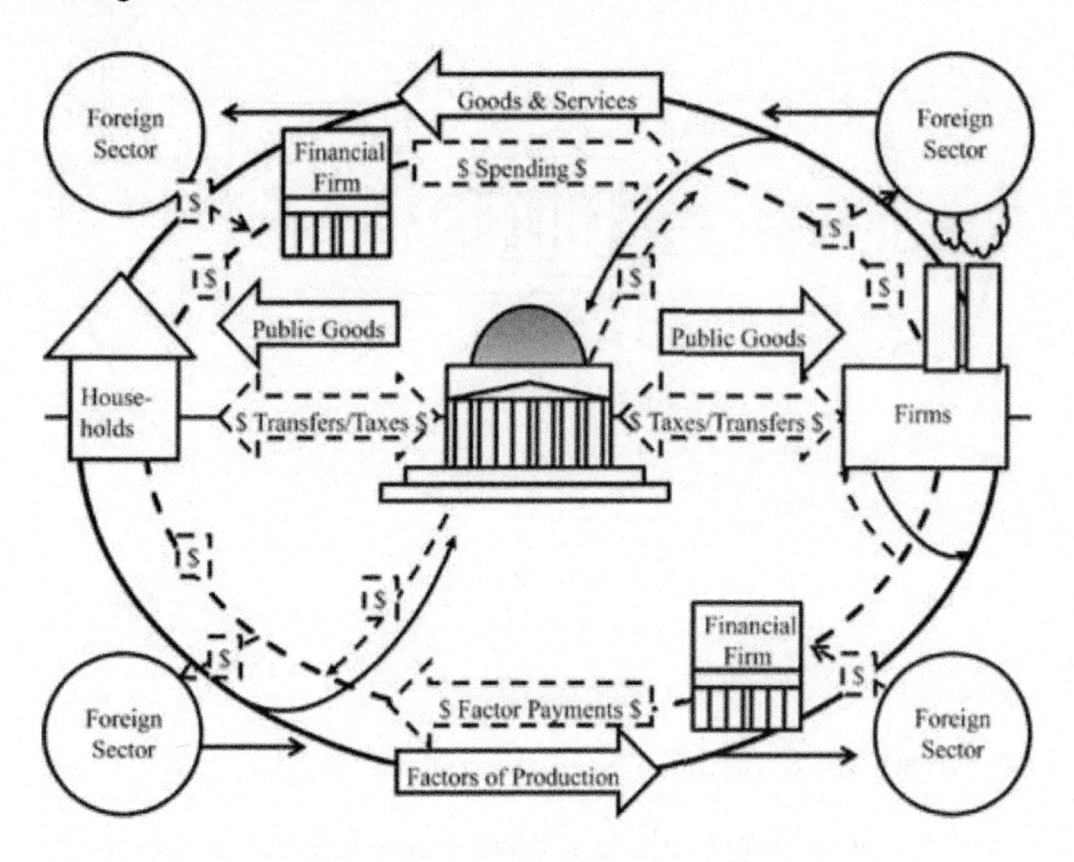

An open economy with financial intermediaries.

The Financial Markets

Up to this point, all of the income that has been earned by households, firms, government, and the rest of the world has been spent in the circular flow model. Private, public, and foreign sectors don't only spend money — they also save. To account for the fact that the different economic sectors save a portion of their income, the circular flow model enters the third dimension.

This third dimension is the **financial markets**. Households save for the future, government can run a budget surplus, businesses retain earnings for future investment, and the foreign sector engages in real and portfolio investment in the United States. Where do all of these savings go? Savings flow to financial intermediaries and from there to all sectors of the economy. When households save, they may buy shares of stock, bonds, or certificates of deposit. Government finances its spending in excess of taxes by issuing the various treasury securities.

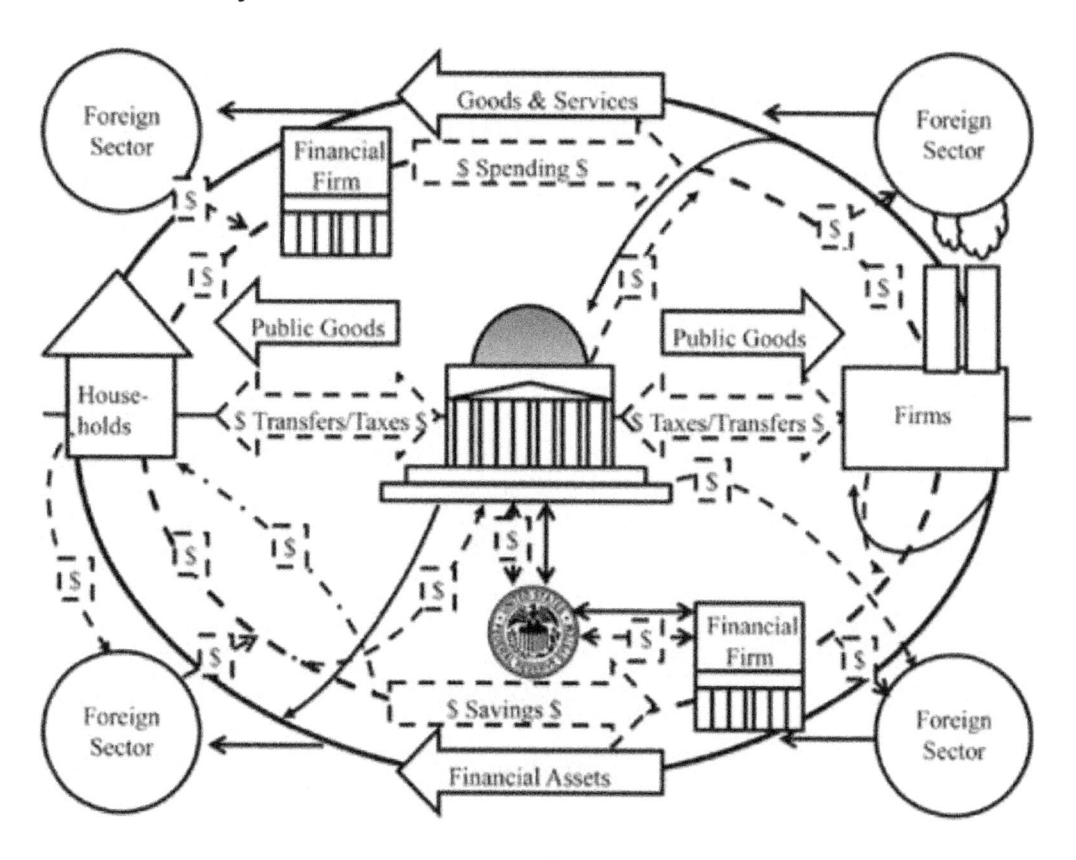

The product market and the financial market.

No circular flow model is complete without a central bank that links government to the financial intermediaries. Careful attention to the diagram reveals that the central bank acts like a heart that pumps money into the economy's circulatory system. Clogged financial institutions, fearful households and firms, and unwilling government can stop an economy. The flow of the real economy in many ways is sensitive to the circular flow of spending and income.

CHAPTER 13: Keeping Score: The Gross Domestic Product

Keeping score is important. If you are trying to lose a few extra pounds, stepping on the scale from time to time allows you to evaluate your performance. In school, teachers assign grades to assess students' level of understanding. In baseball, statistics on nearly every aspect of the game are used to determine the pitchers and the batting order. Data and statistics are useful for making informed decisions. During World War II, the United States government wanted to better understand the economy's ability to generate the necessary materials for the war effort. This led to the development of gross domestic product (GDP) as a means of measuring economic output. GDP is important today as an overall indicator of economic performance.

Wealth and Income

Gross domestic product (GDP) measures the total value of all final production that occurs within a country during the course of a year. GDP is also a measure of annual spending on new domestic production and a measure of income earned from domestic production. To better understand GDP, consider the following example. Assume a simple economy made up of Frank and Louise. Lately, Frank has been complaining about the cold weather, so he offers Louise a dollar to knit a blanket. Louise jumps at the opportunity to make a buck and commences knitting a new blanket. Upon completion, Louise exchanges the blanket for the dollar. In this simple example, what was the value of the economy's spending, income, and output? The answer is one dollar. The one dollar spent by Frank was earned by Louise in exchange for the knitted blanket. In 2009, U.S. GDP was $14.5 trillion, because considerably more than one knitted blanket was produced.

FACT

GDP is not the only measure of economic performance that has been employed. You might remember GNP, or gross national product. The key difference between GDP and GNP is a single preposition. GDP is a measure of all new production that is done *in* a country during the year, while GNP is a measure of all new production that is done *by* a country during the year. Toyotas made in Texas are part of U.S. GDP but not U.S. GNP.

GDP is not static. Instead, GDP is a flow. Imagine a bathtub with a running faucet and open drain. The water flowing from the faucet is GDP, the water accumulating in the tub is the wealth of the nation, and the water exiting the drain represents an outflow like depreciation. As long as the GDP exceeds outflows of wealth, the wealth of the nation grows. Essentially GDP measures the production that is new and is not a measure of accumulated wealth.

A Look Back at Circular Flow

In the circular flow model, the economy has three primary sectors: private, public, and foreign. The private sector is further divided into households and businesses. Each sector contributes to the gross domestic product. Households provide the factors of production that businesses use to produce goods and services. In addition, government provides the public goods not provided by the private sector. Furthermore, the foreign sector acts as a source of the factors of production as well as a source for goods and services. The foreign sector also functions as a market for domestic production.

GDP is represented three ways in the circular flow model. The goods and services that business and government provide represent the value of all domestic production. The spending that the private, public, and foreign sector adds to the circular flow represents total spending. Finally, the rent, wages, interest, and profits earned in the factor market are a nation's income.

Counted or Not?

GDP includes much economic activity but not all of it. As a measure of production, GDP does not include purely financial transactions. When you purchase shares in the stock market, only the broker's commission would be counted in GDP. This is because the purchase of stock represents a transfer of ownership from one shareholder to another, and neither a good nor service is produced. Similarly, transfer payments like Social Security are not computed in GDP. Social Security payments are not made in return for the production of a new good or service but instead represent a transfer from a taxed wage earner to a recipient.

Production for which no financial transaction occurs is also excluded from GDP. A stay-at-home parent who cares for the children, cleans the house, cooks, and runs errands certainly produces something of great value, but because no monetary payment is made, the value is undetermined and excluded. Interestingly, paying someone to do all of the above activities would be included in GDP. Building a deck on your house, mowing your own grass, and changing your own oil are all services that can be purchased, but when you perform them for yourself, they are not included in GDP.

To avoid overstating GDP, resale and intermediate production is excluded. Most home purchases are not counted in GDP. The resale of homes does not represent new production and is excluded. The only time home purchases are included is when the house is newly constructed. The primary reason resale is not counted is to avoid double-counting. Older homes were included in a previous year GDP. Consider the sale of flour, butter, and sugar to a bakery that produces fresh bread. If the purchase of ingredients were included in GDP along with the sale of the fresh bread, the GDP would be overstated. To avoid this, GDP includes only final production of the bread. The price of the bread includes the earlier cost incurred in acquiring the ingredients.

QUESTION

How does GDP account for those who do not rent?

Apartment and house rent is included in consumption, and therefore GDP. Homeowners and mortgage payers do not pay rent, so the BEA imputes a rental payment on their housing. Whether you are an owner or a renter, when it comes to GDP, everyone is a renter.

Illegal production is excluded from GDP because it is difficult to measure. Participants in black market activity have little incentive to provide data for the government to measure production.

Underground activity, like babysitting and lawn care, may go unmeasured because of their cash basis and the failure of the average teenager to properly report income. Some estimates place the size of the illegal and underground economy at approximately 8% of official GDP. If that is the case, then official GDP is understated by more than $1 trillion.

The Spending Approach

GDP can be calculated by adding all of the expenditures on new domestic output. Households, businesses, governments, and other nations all spend money in the economy, and each is represented in GDP by a different spending variable. Each type of expenditure is subject to various influences, and each can reveal important information about economic activity.

Personal Consumption Expenditure

Households engage in personal consumption expenditure. **Personal consumption expenditure**, or simply consumption, is the act of purchasing goods and services. More than two-thirds of all U.S. GDP expenditures fall under the heading of consumption. In other words, Americans shop to the tune of $9 trillion or more per year for new domestic output. The ability of households to consume is chiefly constrained by **disposable income**, which is income after paying taxes. This fact gives government considerable leverage over a household's capacity to consume. When it comes to available disposable income, consumers choose to either spend it in the product market or save it in the financial markets. This means that increased consumption results in less saving, and increased saving leads to decreased consumption.

Consumption is also sensitive to changes in consumers' wealth, interest rates, and expectations of the future. Increases in consumers' wealth tend to encourage consumption, while decreases tend to discourage it. The recent implosion of housing prices has had a negative effect on consumers' wealth. As a result, consumption has declined and saving has increased because consumers are attempting to rebuild their net worth. In addition to housing, much of consumers' wealth is in the form of retirement accounts. Therefore, changes in the stock markets' key indicators, like the Dow Jones Industrial Average and the S&P 500, often indicate the direction of household wealth, and thus consumption.

ALERT

Consumer confidence surveys like those conducted by the University of Michigan and the Conference Board provide economists with an idea of how consumers "feel" about the future. Increases in consumer confidence often lead to increased consumption, and ultimately increased GDP and lower unemployment.

Durable goods like cars, appliances, and furniture are often purchased using borrowed money. Interest rates are then part of the equation to consume durables. High interest rates discourage durable goods consumption, and lower interest rates tend to encourage the purchase of durable goods. Plans designed to encourage durable goods consumption include 0% financing, interest-free for three years, and no payments until the next year. Expectations as a determinant of consumption are easily understood. When people fear the future, they tend to save and not make any major purchases. In contrast, people that are upbeat about the future are more inclined to shop and less inclined to save.

Economists recognize that some consumption is independent of disposable income and refer to this as **autonomous consumption**. During periods of recession, consumption of durable goods is curtailed while consumption of essentials, like food, rent, clothing, and health care, remains relatively unaffected. Do you remember what it was like on September 12, 2001? What type of consumption occurred and what type did not? Chances are that autonomous consumption of necessities and nondurables continued unabated, but consumption of durable goods (except for guns and ammo) and discretionary spending came to a halt.

Gross Private Investment

Households and businesses engage in another expenditure referred to as gross private investment. **Gross private investment** includes business purchases of capital and inventory and household purchases of new homes. Gross private investment is further broken down into net investment and depreciation. **Net investment** is the purchase of new capital that expands the economy's productive capacity, whereas **depreciation** is investment spending to replace worn capital. It is important to note what investment is not. When it comes to calculating GDP, investment is not the purchase of stocks and bonds. That is financial investment or saving. What is being considered here is **real investment**. However, this real investment in the economy is financed by the financial investment that occurs in financial markets.

FACT

Remember that households either spend or save their disposable income. An important equation in economics states that savings equals investment. The money that households don't spend is the same money that businesses use to invest. Because savings are able to flow in and out of the country, increased domestic saving does not necessarily translate into an equal increase in domestic investment.

The level of investment in the economy is influenced by expectations of future business conditions and interest rates. Positive expectations tend to boost investment while negative expectations result in less investment. Businesses respond this way in order to have the right amount of productive capacity to meet the expected future demand for their products. Interest rates are also a major consideration in the decision to invest. As interest rates rise, the relative profitability of investment decreases. Decisions to invest compare the expected rate of return to the current interest rate. As long as the expected rate of return exceeds the interest rate, businesses will undertake investment with the expectation of profit. Increased interest rates have the effect of making fewer investments profitable.

Investment is divided into planned and unplanned investment. During periods of economic "normalcy," businesses invest in capital and in inventory in order to turn around and sell it at a profit. This planned investment assumes that business conditions will continue according to the producers' expectations. When unplanned investment occurs it is a sign of bad things to come. Auto dealers order inventory from the factories to sell to consumers. If inventory begins to accumulate, that means consumers are not buying. The dealers stop ordering and the factory stops production. If this halt in production is pervasive, mass unemployment and economic recession occurs.

Government Spending

Government spending includes federal, state, and local expenditures on capital, infrastructure, and employee compensation. Military expenditures, road construction, and teacher salaries are all included. Government spending is financed by taxation and borrowing. The opportunity cost of government spending is the forgone consumption and gross private investment that might have otherwise occurred. Transfer payments are not included as government spending for the purpose of calculating GDP.

Government spending is limited by the amount of tax revenue and the ability to borrow. For a country like the United States this is not much of a limiting factor, as taxes are mandatory and U.S. government debt is a popular vehicle for financial investment. Regardless of the political party in power, government spending tends to have an upward trajectory. In the United States, all politics is local. This means that representatives and senators have a strong incentive to direct federal spending towards their respective districts and states in order to win support for re-election. Failure of politicians to bring home the pork is usually cause for a new politician to replace the incumbent.

Government spending is often used to stimulate consumption during recessions because government often has the ability and will to spend when other sectors of the economy are not. The effectiveness of this spending is hotly debated among the different schools of economic thought. Those economists influenced by the work of John Maynard Keynes tend to support government spending as a stimulus to the economy. Economists with a classical or libertarian bias often argue that government spending offsets more efficient private spending and should not be used as a tool of stimulus.

Net Exports

Net exports, or exports minus imports, are the last spending variable in measuring GDP. Export of new domestic production adds to GDP. Imports, on the other hand, subtract from GDP. The United States typically runs a balance of trade deficit, so in most years net exports are deducted from, instead of added to, GDP. In recent years, export growth has helped to sustain GDP. Compared to most other nations, U.S. net exports represent a very small percentage of economic activity. Even though no country engages in as much international trade by volume as the United States, the United States remains one of the least trade-dependent nations in the world.

FACT

Net exports may not be a real big deal in America, but for developing countries net exports are the road to economic growth. China is a key example of a country dependent on net exports for growth. As the Chinese economy matures, its industry may become more oriented toward domestic production.

Net exports are influenced by exchange rates, and like consumption and investment, also by interest rates. Appreciation of the dollar makes American goods relatively expensive, so exports decline and imports rise. Depreciation of the dollar, on the other hand, makes American goods relatively cheap, so exports increase and imports decrease. Interest rates impact net exports through their effect on the exchange rate. High interest rates in America lead to an appreciation of the dollar, which reduces net exports, but low American interest rates help to depreciate the dollar and encourage net exports.

Income Approach to GDP

The ability to engage in consumption, investment, government spending, and net exports derives from the income earned producing domestic output. Again, income includes all of the rent, wages, interest, and profits earned by selling the factors of production in the factor market. Measuring income is more complex than measuring spending, and this requires some mental gymnastics on the part of economists. One reason is that profits flow to corporations, shareholders, and proprietors. Also, taxes and subsidies distort the difference between the price paid in the market and the income earned by producers. In the end, measuring income is a bit more complex for economists than measuring expenditures, but for now it is sufficient to conclude that income equals the sum of rent, wages, interest, and profits. Let the economists at the BEA fret over the details. One last practical problem arises when measuring income versus spending. Producers and households have an incentive to underreport income in order to reduce their tax liability.

Nominal Versus Real

The concepts of nominal and real appear throughout economics, and GDP is no exception. **Nominal GDP** is reported using current prices. In order for economists to make valid comparisons in GDP from year to year, the price changes that occur with time's passage have to be addressed. **Real GDP** reports output, holding prices constant. If the change in prices (inflation) is not accounted for in calculating GDP, results may be misleading.

Nominal GDP must be deflated in order to calculate real GDP. Assume an extremely simple economy that produces multicolored beach balls. In 2009, the economy produced 100 beach balls that were all purchased by consumers at $1.00 a piece. In 2010, the economy produced 100 identical beach balls that were all purchased by consumers for $1.25 a piece. Given that information, nominal GDP for each year can be calculated by multiplying the number of beach balls by that year's current price, so in 2009 nominal GDP was $100 and in 2010 it was $125. An outside observer might come to the incorrect conclusion that output increased by 25%. The reality was that output did not change, but prices rose by 25%. To compare what *really* happened, prices must be held constant. Using 2009 prices, the *real* GDP for 2009 and 2010 is $100, in other words real output remained constant.

Real GDP Changes and the Business Cycle

Over the last fifty years, America has undergone alternating periods of recession and economic expansion. The ups and downs have occurred against a background of long-run economic growth. Since 1960, the U.S. real GDP has increased by more than $10 trillion. Economists refer to this series of expansions and contractions as the business cycle. Different theories have been suggested to explain the business cycle.

As the workforce and productivity grow, so does the economy's capacity to produce. This explains why the economy has grown over time. The periods of expansion and contraction are attributed to differences between total spending on output relative to the economy's long-term capacity to produce. During periods of expansion, spending increases to the point where the economy exceeds its long-run production capacity. Contractions occur as total spending decreases and excess productive capacity remains.

If you have ever been on a long car trip, you can understand the business cycle. Imagine cruising along a two-lane highway, going with the flow of traffic. This going with the flow is the norm and represents the average rate of economic growth. Occasionally, somebody will be moving a little too slowly, so you scan for oncoming traffic. If it's clear, you accelerate and pass the slower driver. Passing represents those periods where spending exceeds the productive capacity. Every once in a while, you make a mistake while trying to pass slower traffic. You scan for oncoming traffic and make your move, only to discover that a fully loaded eighteen-wheeler is barreling down the highway in the oncoming lane. You immediately switch back into your lane, shaking violently, and slow to thirty miles per hour, thankful that you are still alive. Events like this are representative of economic contractions. Eventually, you compose yourself and start to travel with the flow of traffic again.

Now, to really understand the business cycle, imagine that you are blind-folded the whole time and relying on an extremely myopic passenger speaking Swahili to provide information about what is happening. To be sure, it would be an interesting ride. Why the blindfold and Swahili? Outside of psychics, prophets, meteorologists, and the rare economist, most Americans have difficulty seeing into the future. The language of economists and financial experts is often difficult to interpret. Keynes explained the business cycle as being caused by "animal spirits" where businesses are in effect bipolar. Keynes's animal spirits represent the emotion that clouds rational decision-making. These animal spirits are expressed through businesses' willingness to invest. When businesses are manic they invest heavily, only to fall into stages of depression where they are unwilling to invest. Milton Friedman's monetarists explain the business cycle as being caused by poor management of the money supply. Periods of overexpansion are produced by too much money, and periods of contraction are caused by too little money in circulation. Most other theories explain the business cycle as ultimately being caused by spending changes, but one referred to as the Real Business Cycle Theory focuses on changes in productivity as being the ultimate cause of the cycle.

What GDP Doesn't Tell Us

In a 1968 speech, the late presidential candidate Robert F. Kennedy stated the following about the weaknesses of our key measure of economic performance at that time, gross national product. The same weaknesses apply to GDP:

> *"Too much and too long, we seem to have surrendered community excellence and community values in the mere accumulation of material things. Our gross national product — if we should judge America by that — counts air pollution and cigarette advertising, and ambulances to clear our highways of carnage. It counts special locks for our doors and the jails for those who break them. It counts the destruction of our redwoods and the loss of our natural wonder in chaotic sprawl. It counts napalm and the cost of a nuclear warhead, and armored cars for police who fight riots in our streets. It counts Whitman's rifle and Speck's knife, and the television programs which glorify violence in order to sell toys to our children.*
>
> *"Yet the gross national product does not allow for the health of our children, the quality of their education, or the*

joy of their play. It does not include the beauty of our poetry or the strength of our marriages; the intelligence of our public debate or the integrity of our public officials. It measures neither our wit nor our courage; neither our wisdom nor our learning; neither our compassion nor our devotion to our country; it measures everything, in short, except that which makes life worthwhile. And it tells us everything about America except why we are proud that we are Americans."

Kennedy's point should be clear. GDP, for all of its inclusiveness, excludes many important things. However, consider the fact that as real GDP has increased, the burdens of scarcity and the incidence of absolute poverty have been lifted for millions of people. Yesterday's relative wealth is today's relative poverty. Compared to the lives of Americans of previous generations, the availability of health care, education, nutrition, sanitation, and housing has increased with the increases in real GDP. These have led to an increase in longevity.

QUESTION

What is GDP per capita?

GDP per capita is the GDP divided by the population. As an indicator of overall well-being, it is subject to a major flaw. GDP per capita gives no indication as to how income is distributed among the population. The United States and Norway both have high GDPs per capita, but the key difference is that U.S. incomes are unequally distributed compared to Norway. High GDP per capita does not necessarily mean that there are not those in society living in relative poverty.

The increase in real GDP has been accompanied by more leisure time as well. The average work week has steadily declined and the average number of vacation days has increased. As a measure of well-being, the GDP has both strengths and weaknesses.

Critics of GDP say that it does not take into account environmental degradation. Because GDP is focused on spending and output, it creates an incentive to pursue greater amounts of production in order for growth to continue. This growth can come at the cost of the environment. Deforestation, climate change, pollution, and other environmental ills are, according to critics of the measure, the logical outcome of this narrowly sighted focus on GDP. Others defend GDP, stating that it is because of the increase in real GDP that people are wealthy enough and have the time to care for the environment. Today, the countries with the highest real GDPs are often the very ones that are doing the most to address the issues that environmentally conscious citizens raise.

CHAPTER 14: Where Did My Job Go?

One of the most gratifying things that you can hear is, "We like you, and you're the right fit for this company. Congratulations, you're hired!" One of the worst things anyone can hear is, "You're fired." Unemployment can make economics suddenly appear very relevant to your life. Economists define, measure, classify, evaluate, and seek to understand this all-too-common phenomenon. Many economists have made it their life's work to minimize the problem of unemployment, and policymakers are under political pressure to do so as well.

What Unemployment Is and What It Is Not

According to the Census Bureau, the 2008 U.S. population was approximately 300 million, of which 145 million were employed. How many were unemployed? It might come as a surprise to you that the answer to that question cannot be determined from the information given. True, you can infer that 155 million did not work, but that does not necessarily mean that they were all unemployed. Are toddlers and kindergarteners unemployed? Of course they are not. To determine the number of unemployed, you must first define the term unemployment.

Persons sixteen years of age or older are considered **unemployed** if they have actively searched for work in the last four weeks but are not currently employed. The **employed** are those who have worked at least one hour in the previous two weeks. People who meet neither criterion are not considered in the **labor force**, which is the number of employed persons plus the number of unemployed persons. The **unemployment rate** that you hear quoted in the news is not a percentage of the population, but a percentage of the labor force that is not currently employed.

QUESTION

Does an increase in the unemployment rate mean that fewer people are employed?

Not necessarily. It is possible for the unemployment rate to increase at the same time the number of employed is increasing. A demographic shift in the labor force like the entrance of women during World War II or the return to civilian life of members of the military can create a condition where the ratio of unemployed to the labor force increases even while employment is increasing.

There are many reasons for not participating in the labor force. Full-time students, retirees, stay-at-home parents, the disabled, and the institutionalized do not participate. Members of the military on active duty are not considered part of the labor force either. At any point in time there are people entering, exiting, and re-entering the labor force. Furthermore, people are forever getting hired, fired, and furloughed. They are also quitting, cutting back, and retiring. The labor force is in constant flux, which makes measuring unemployment a daunting task.

Measuring Unemployment

The U.S. Department of Labor's Bureau of Labor Statistics (BLS) monitors unemployment in the United States. Once a month, the Census Bureau conducts the Current Population Survey. Approximately 60,000 sample households are briefly questioned about their participation in the labor force. The BLS uses the data from the survey to calculate the various employment statistics used by economists and policymakers.

FACT

The latest unemployment and labor force data can be found at *www.bls.gov*. The Bureau of Labor Statistics website allows you to access hundreds of tables, graphs, press releases, and research articles. In addition to unemployment data, the website also contains the latest information on inflation.

In addition to the population survey, economists look at payroll employment records, new claims for unemployment insurance, and other data to get a complete picture of the country's unemployment. Nonfarm payroll employment records give economists a good idea of how much hiring is taking place in the economy. New claims for unemployment insurance act as a verification of the current population survey's results. Policymakers also look at the weekly hours worked in manufacturing. Declines in the weekly hours worked indicates that factories are idling back and may be laying off workers. Increases in weekly hours worked may indicate that firms will hire in the future.

In addition to calculating unemployment, the BLS uses the survey data to calculate the labor force participation rate and the employment-to-population ratio. The **labor force participation rate** is the percentage of the working age population classified as either employed or unemployed. In the United States, the average labor force participation rate is about 65%. The **employment-to-population ratio** is the percentage of the working age population that is classified as employed. Both ratios have declined over the last ten years as numbers of young people have delayed entry into the labor force. This may be as a result of increased college or university enrollment and increases in military enlistment.

Falling Through the Cracks

Because of unemployment's narrow definition, many people that you might consider unemployed or underemployed are not captured in the official unemployment statistic. **Marginally attached workers** are not considered in the official unemployment rate. These are people ready and available to work, who have conducted a job search within the past twelve months, but have not searched in the last four weeks. Some people have given up the job search in frustration. These **"discouraged workers"** are unemployed in the general sense, but because they do not meet the technical definition, the official unemployment rate does not reflect their numbers. In addition, many full-time workers who have lost jobs have been rehired as part-time workers. The automotive engineer in Detroit who now works the drive-through lane at a fast food restaurant is an example of this.

The BLS publishes several unemployment rates in addition to the official unemployment rate (**U3**) to measure those who fall between the cracks of the official rate. **U1** only includes people unemployed fifteen weeks or longer. **U2** only includes people who have lost a job as opposed to those who have quit or those who have entered or re-entered the labor force. **U4** adds discouraged workers to the official unemployment rate. **U5** includes all marginally attached workers. Finally, **U6** is the most all-inclusive measure of unemployment and includes all of the above plus those who are employed part-time because of economic reasons.

Types of Unemployment

Economists make qualitative distinctions in the reasons for various classifications of unemployment. Not all unemployment is the same. Some types are actually positive for the individual and the economy. Other types are bad for the individual but benefit society. Last, there is one type of unemployment that is both bad for the individual and is costly to society. The three types of unemployment are frictional, structural, and cyclical.

Frictional Unemployment

Is 0% unemployment a good goal for society? It is most definitely not. A 0% unemployment goal ignores the presence of frictional unemployment. **Frictional unemployment** occurs when people voluntarily enter the labor force, or when they are between jobs for which they are qualified. It is frictional because the labor market does not automatically match up all available jobs with all available workers. Instead, job search requires time for the right worker to find the right job. Both workers and society benefit when job applicants are matched to the appropriate job. You want mechanical engineers to get jobs in mechanical engineering, not pet grooming.

The rate of frictional unemployment is relatively low, and as technology increases and search times diminish, it becomes even lower. The advent of online job search sites and social networking has reduced job search times for many workers. Government incentives create variations in frictional unemployment rates between countries. Generous unemployment benefits give workers an incentive to spend more time in searching for a job and thus increase the rate of frictional unemployment for the country. Compared to Europe, American unemployment benefits are less generous. As a result, Americans spend less time searching for jobs and the rate of frictional unemployment is relatively lower.

Structural Unemployment

Structural unemployment occurs when job seekers' skill sets are not in demand because of geography or obsolescence. As industries die out in certain regions of the country or relocate to other regions, the workers may not be able to move with the job. This leaves workers with a skill set that is no longer in demand. These workers must either retrain or accept a lower-paying job in an industry that requires less skill. Structural unemployment is often the outcome of what the economist Joseph Schumpeter called **creative destruction**. As innovation occurs, old technologies and industries are destroyed, which frees up the resources for the new technology and its industry.

The invention of the personal computer was the death knell for the typewriter. As the new technology advanced, the old technology and its industry were destroyed. Over time, skilled typewriter repair technicians found that their skill set was no longer in demand and faced the permanent destruction of their jobs. As this was occurring, new jobs were being created in the new industry. The problem for workers is that their skill sets may not translate into the new industry. The solution for structural unemployment is education and retraining.

QUESTION

Is it possible for a person to be both frictionally and structurally unemployed?

The classification of unemployment is not an exact science. A person who voluntarily leaves a job to search for another job for which he is unqualified would be both frictionally and structurally unemployed. Throw in a recession and this person might find himself frictionally, structurally, and cyclically unemployed.

Another reason for the presence of structural unemployment is the presence of efficiency wages in the labor market. **Efficiency wages** are those that exceed the equilibrium market wage. The purpose of efficiency wages is to encourage worker productivity. Employees who earn efficiency wages know that they are unable to earn equivalent wages with competing firms, so they are motivated to produce more output for the paying firm. The efficiency wages have the added effect of enticing more people to enter the labor market. Because these entrants are attracted to the efficiency wage and not willing to work for the lower market wage, they represent an increase in the level of structural unemployment. As more entrants are attracted to the market, less-skilled workers face more competition for jobs and suffer higher rates of unemployment.

Cyclical Unemployment

The most insidious type of unemployment is cyclical. **Cyclical unemployment** occurs because of contractions in the business cycle. It is not voluntary, nor is it the result of a skill-set mismatch. During periods of recession, the official unemployment rate increases as cyclical unemployment adds to the always-present frictional and structural rates of unemployment. The recession that began in 2007 saw the official unemployment rate increase from 5% to 10%. The additional 5% increase is directly attributed to cyclical unemployment.

The real problem with cyclical unemployment is that it creates a feedback loop. As one group becomes cyclically unemployed, they cut back on spending, which leads to more cyclical unemployment. This feedback loop resulted in 25% unemployment during the Great Depression. Policymakers respond to cyclical unemployment with discretionary fiscal and monetary policy. In addition, automatic stabilizers like unemployment compensation help to dampen the feedback loop by allowing affected workers to have some capacity for spending. Ultimately the goal of policymakers is to eliminate cyclical unemployment altogether.

Full Employment

When the economy is producing at its optimum capacity, cruising down the road at the speed limit, neither speeding nor driving too slowly, it is safe to assume that the economy is also at full employment. **Full employment** occurs when cyclical unemployment is not present in the economy. This economic nirvana is the goal that policymakers seek to maintain.

Economists associate full employment with the **natural rate of unemployment**. The natural rate hypothesis advanced by Nobel economists Milton Friedman and Edmund Phelps suggests that in the long run there is a level of unemployment that the economy maintains independent of the inflation rate. The idea is that left alone, the economy will maintain full employment and experience the natural rate of unemployment most of the time.

It is possible for the natural rate of unemployment to vary. If frictional or structural rates of unemployment were to change, then the natural rate of unemployment would change. A technology that permanently reduces search time for jobseekers would have the effect of both reducing frictional unemployment and the natural rate of unemployment. Permanent changes in unemployment compensation that encourage or discourage lengthy periods of unemployment would also affect the natural rate. Finally, lasting increases in worker productivity would reduce the natural rate of unemployment.

Different countries have different natural rates of unemployment. Economies that are more market oriented like the United States have low natural rates of unemployment. Socialist economies tend to have higher natural rates of unemployment. Economists theorize that socialist economies have both higher frictional and structural rates of unemployment because of government policies that lead to less-flexible labor markets. The highest natural rates of unemployment occur in countries where labor is unskilled and immobile, and job creation is stymied by corrupt, inefficient governments.

Why Unemployment Is Bad

Unemployment creates a measurable cost for the economy and individuals.

The opportunity cost of unemployment is immense when considering the scale of the United States economy. When workers are unemployed, they are unable to produce output. According to the economist Arthur Okun, for every 1% that the official unemployment rate exceeds the natural rate of unemployment, there is a 2% gap between actual and potential real GDP. Given the GDP and unemployment figures from 2009, when actual output was $14 trillion and unemployment was 10%, and assuming a natural rate of 5%, actual output may have been $1 trillion to $2 trillion below its potential. By way of comparison, a $2 trillion output gap is like sacrificing the entire economic output of France.

FACT

Incentives matter. A string of bus bombings in Israel drew worldwide attention when it was discovered that the families of suicide bombers were being paid upwards of $30,000 by Iran. Unemployment rates are extremely high in the West Bank, and for some, the money made by killing themselves was greater than what could be made given severely limited employment opportunities.

The costs to the individual are heavy as well. An extended period of unemployment can wipe out a family's personal savings and leave them in debt. Unemployment disrupts the normal flow of life and if prolonged can possibly lead to health and psychological problems for affected individuals. Also, the incidence of family violence is directly related to changes in the unemployment rate. Furthermore, periods of high unemployment are also associated with increases in the divorce rate and child abandonment.

Prolonged, pervasive unemployment is directly linked to crime and civil unrest. Areas plagued with persistent high unemployment are also plagued with both violent and property crime. A trip to America's inner cities provides the anecdotal evidence for this. Much of the unrest in the developing world coincides with high rates of unemployment. It is a very rare day when someone takes the day off of work to riot or blow something up. Unemployment, it seems, creates the necessary condition for many of the world's problems.

Trends and Demographics

The United States economy has undergone several major shifts. Initially, the United States was an agrarian nation and most employment was related to farming. The industrial revolution saw a shift toward manufacturing employment. Today, most jobs are created in the service sector. As America moves away from agriculture and manufacturing, fewer and fewer jobs in those areas will remain. Globalization has shifted many low-skill jobs overseas, which leaves America's unskilled workers with fewer opportunities.

Demand for labor is driven by worker productivity. The more skills workers possess, the greater the demand for their labor. To stay competitive, future workers must realize that they are not just competing against their fellow Americans but against the rest of the world. The day when you could graduate from high school and get a good paying job at the factory are gone, unless, of course, you live in China. To compete in the global job market, Americans must be willing to train, stay mobile, and constantly adjust to the changing needs of their employers.

When it comes to the demographics of unemployment, the facts show significant differences between subpopulations. Men have higher rates of unemployment than women. Whites have lower rates of unemployment than Hispanics. Hispanics have lower rates of unemployment than African Americans. Younger workers are more likely to be unemployed than older workers. Educated workers are far less likely to be unemployed than dropouts. To demonstrate the variance in unemployment, consider the difference in 2008 unemployment rates between married, white females over age 25 and single, black males under age 25. The rate for the first group was 3.3% while the rate for the second group was 17.5%. The most important demographic determinant of unemployment, however, is educational attainment. Even during recessions, unemployment rates for college graduates remain below 6%.

CHAPTER 15: Inflation

Were you a little frustrated a few years back when gas prices suddenly rose? The increase in gas prices probably created some hardship as you altered your spending in order to accommodate its higher cost. Now imagine that not just gas prices but the price of almost everything you buy suddenly and unexpectedly increases. If you are on a fixed income, then there is only so much altering you can do to a budget before you realize that high prices are killing your finances. Inflation is a phenomenon that you need to understand if you want to comprehend how the economy works.

What Is Inflation?

No word strikes more fear into the hearts of central bankers than inflation. Defined as a general increase in prices or as a decrease in money's purchasing power, inflation creates problems for more than central bankers. Inflation affects everyone in the economy. Governments, businesses, and households are subject to inflation's influence. Inflation is either created by excessive demand or increases in producers' per unit costs, but it is sustained by too much money in circulation. Left unchecked, inflation can have cataclysmic results for a society.

During the 1920s, the Weimar Republic of Germany suffered from extreme inflation. Instead of taxing or borrowing to raise revenue, the government began to print money for the purpose of making its purchases. The result was runaway inflation. Some historians credit the period of inflation and the resulting loss of confidence in the

Weimar Republic for sowing the seeds of Hitler's eventual rise to power.

FACT

Hungarian inflation was so severe in 1945 and 1946 that prices were measured in not tens or hundreds but billions, trillions, and even octillions. By the end of the period of Hungarian hyperinflation, the total supply of pengos in circulation had less value than a single pengo did at the beginning.

If you were around in the 1970s, then you might remember the time period known as the great inflation. The Vietnam War, OPEC, the collapse of the adjustable peg system (where world currencies were pegged to the dollar), and poorly managed monetary policy created conditions of rising prices and uncertainty. Although American inflation did not even come close to approaching the levels of the Weimar Republic and modern-day Zimbabwe, it was enough to cause political turmoil and bring lasting change to the way policymakers manage the price level.

Measuring Inflation

Inflation is the rate of increase in the average price level of the economy. To measure inflation first requires that the price level be measured. Economists have come up with different ways to measure the general price level in the economy, and therefore, inflation. The most often cited measure of inflation is the change in the **consumer price index** (CPI). In addition, economists and policymakers also pay attention to changes in the **producer price index** (PPI) and the **personal consumption expenditure deflator** (PCE deflator).

ESSENTIAL

Economists and policymakers pay careful attention to inflation. In order to make appropriate policy though, they must choose to ignore certain types of inflation. The BLS and BEA both publish headline and core inflation measures. Headline CPI includes the entire market basket, while core CPI excludes more volatile food and energy prices. The Fed pays attention to core PCE when making its policy decisions. You do not want to raise interest rates just because corn and gas prices increased.

The CPI is a market basket approach to measuring the price level and inflation published by the BLS. The CPI measures the average cost of food, clothing, shelter, energy, transportation, and health care that the average urban consumer buys. To understand CPI, imagine that you are given a shopping list of thousands of different items. You are then told to research and write down the price of each specific item, and afterward add them all together. The total cost of the list would represent an average price level. Further assume that a year later you took the same list and repeated the process. Increases in the shopping list's total would represent inflation.

The PPI is similar to CPI, but instead of consumer prices, PPI looks at producer prices. The PPI includes all domestic production of goods and services. Unlike CPI, the PPI also includes the prices of goods sold by one producer to another. Changes in PPI can be used as a predictor of future changes in CPI. Before consumer prices change, the producer price changes. Because it predicts changes in CPI, the government and central bank use the PPI to create fiscal and monetary policy in anticipation of possible consumer inflation.

FACT

In order to remove the effects of inflation from nominal GDP to arrive at real GDP, the BEA also computes the GDP deflator. This price index is inclusive of all final domestic prices. Where the CPI measures groceries, the GDP deflator measures the price of groceries and the price of armored personnel carriers. This broad measure is not of much use to individual consumers, but it is important for understanding the whole economy.

The PCE deflator is a broad measure of consumer inflation published by the BEA. Unlike the CPI, which measures a fixed market basket, PCE deflator measures all of the goods and services consumed by households and nonprofit institutions. PCE deflator is a more comprehensive measure. Because it does not deal with a fixed market basket, it better reflects consumers' tendency to substitute more-expensive items with less-expensive items and their tendency to vary consumption as time passes.

Types of Inflation

There are two primary types of inflation: demand-pull and cost-push. Understanding which type of inflation is occurring at any given point in time is important if policymakers want to respond appropriately. The two types of inflation are not mutually exclusive, so it is possible for both to occur simultaneously. Left untreated, inflation can cause a wage-price spiral or even hyperinflation.

Demand-Pull Inflation

Demand-pull inflation occurs when spending on goods and services drives up prices. Demand-pull inflation is fueled by income, so efforts to stop it involve reducing consumer's income or giving consumers more incentive to save than to spend.

Demand-pull inflation persists if the public or foreign sector reinforces it. Low taxes and profligate government spending exacerbate demand-pull inflation. A failure of the central bank to reign in the money supply also makes the demand-pull inflation worse. Demand-pull inflation can spread across borders as well. China and India's economic growth not only puts pressure on prices in these countries but also on prices worldwide as the demand for imports increase.

If government spending is financed by printing currency or by the central bank monetizing the debt, demand-pull inflation can become hyperinflation. **Hyperinflation** is defined as annual inflation of 100% or greater. All cases of hyperinflation have been accompanied by the government or central bank issuing too much money.

QUESTION

What does "monetizing the debt" mean?

Monetizing the debt refers to the process by which the central bank buys new government debt, thus increasing the supply of money in circulation. When debt is monetized, the government is able to spend without raising taxes or borrowing from the private sector. The downside is that debt monetization is extremely inflationary.

Cost-Push Inflation

Cost-push inflation occurs when the price of inputs increases. Businesses must acquire raw materials, labor, energy, and capital to operate. If the price of these were to rise, it would reduce the ability of producers to generate output because their unit cost of production had increased. If these increases in production cost are relatively large and pervasive, the effect is to simultaneously create higher inflation, reduce real GDP, and increase the unemployment rate. You might recognize this combination by another name, **stagflation**. In the 1970s, OPEC cut oil production, which led to much higher energy prices along with double-digit inflation and unemployment. Because producers faced higher operating costs, they reduced output. Relative to the demand for their products, the supply decreased, which resulted in cost-push inflation.

If cost-push inflation has a bright side, it is the fact that it is self-limiting. Cost-push inflation is associated with decreases in GDP. The decreased GDP and resulting high unemployment helps to bring producer prices back down. The trick to combating cost-push inflation is realizing that it is not demand-pull. The policy prescription for each is different, and applying the wrong prescription can create more problems than it solves. It is the unemployment issue that usually spurs policymakers to action. If they respond to the increased unemployment by increasing spending, the inflation problem is made worse. A **wage-price spiral** can result if the policy responses create more demand for goods and services at the same time that unit costs are rising. By way of analogy, the prescription for a grease fire is different from that of a forest fire. Grease fires are put out by removing the source of oxygen, while a forest fire is extinguished with water. If you pour water on a grease fire, then things only get worse. This is what happened in the 1970s. Instead of letting cost-push inflation run its natural course, the Fed poured money on it, and inflation worsened.

Who Gains and Who Loses from Inflation?

Inflation creates winners and losers. Knowing who wins is important for understanding why it is sometimes allowed to persist. When inflation is expected and stable, it is rather benign. People and institutions can plan for it and build it into their decision-making. If inflation is unexpected, it creates a win-lose situation in society. Who stands to gain from inflation?

Benefiting from Inflation

First consider what inflation is: a general increase in prices and a corresponding decrease in money's purchasing power. Borrowers benefit from a general increase in prices or a reduction in purchasing power. When individuals, businesses, and governments borrow, it is usually at a fixed rate of interest that had some expected level of inflation built into it. If higher than expected inflation occurs, then the real value of the borrower's debt is reduced. Assume that banks lend billions of dollars at a fixed nominal interest rate of 5%. If inflation were to unexpectedly increase from 2% to 4%, then borrowers' real interest rate paid would be reduced from 3% to 1%. In simpler terms, the money that was lent was more precious than the money being repaid.

Another group that benefits from an increase in consumer prices in the short run is producers. When unexpected inflation occurs, consumer prices rise while wages paid to employees remain relatively stable. This allows producers to experience higher profits for a time until wages adjust to reflect the higher prices consumers are paying.

In the past, many governments in the developing world tried erasing their foreign debts by overprinting their currency. Faced with much external debt, governments would devalue their currency in order to satisfy the debt. Given the current size of the United States debt, some fear that the American government might be tempted to do something similar. Most developed nations have independent central banks to act as a check on government's incentive to overprint currency. In the United States, the Federal Reserve is somewhat insulated from political pressure and can constrain the money supply when government's incentives are to expand it.

Losing with Inflation

Inflation harms more than helps. Lenders and savers both lose when inflation exceeds expectations. Both earn interest rates that assume some rate of inflation, and when the actual rate exceeds the expected rate, savers and lenders are harmed. Maybe you save money in a bank CD. Assume you purchase a $1,000 one-year simple CD that pays 4% nominal interest. If inflation increases unexpectedly from 2% to 5%, then the real interest rate you earn is approximately -1%. You are worse off than when you started. In nominal terms you still made $40.00 of interest. The problem is that the $1040 that you now have has less purchasing power than the $1000 you started with.

Inflation is thought to be harder on those with lower incomes. People with low incomes tend to have more of their wealth in the form of cash than do those with higher incomes. High-income earners have cash to be sure, but they also are more likely to have much of their wealth in other real and financial assets. For the poor, inflation exacts a heavier toll because it destroys the value of their chief asset, which is cash. The higher-income earners are able to offset some of inflation's effect by holding assets that actually appreciate with inflation.

Those living on fixed incomes are also harmed by inflation. During periods of unanticipated inflation, fixed-income earners see their real incomes decline. Professionals on a fixed salary or retirees living on a fixed pension lose purchasing power as long as the rate of inflation exceeds the rate at which their pay increases. To mitigate some of these effects, employers, or in the case of social security recipients, the government, will adjust pay to inflation through the use of **cost of living adjustments** (COLA). Even with COLA, fixed-income earners are still harmed by inflation as the cost adjustment lags inflation. During periods of higher than expected inflation, fixed-income earners are forever playing a game where their pay increases are too little and come too late.

ESSENTIAL

Next time you sit down at a restaurant with laminated menus, consider what that says about inflation. When restaurants laminate their menus they are not only protecting them from spills, but they are also testifying that prices are relatively stable. When inflation is out of control, restaurants must continually update their prices, so they do not want to fix prices on the menu. Unlike most restaurant prices, seafood prices are highly volatile; that is why they are usually priced on a chalkboard.

Inflation creates practical problems for individuals and businesses in an economy. Because money is quickly losing value, consumers must engage in transactions more frequently as they rush to spend whatever money they have. The increase in transactions creates what is called **shoe-leather costs**. You wear out your shoes quicker when inflation is present because of the increase in your transactions. Inflation also poses a problem for producers who constantly have to reprice their goods as inflation continues. Remember that there is no such thing as a free lunch. Placing prices on goods is not free. If inflation is high, then significant costs are created as businesses pay employees to reprice their items. Persistent inflation results in forgone output as labor resources are put to the task of keeping up with ever-changing prices.

Disinflation

One of the triumphs of Fed policy came in the early 1980s when the central bank under the direction of then chairman Paul Volcker raised interest rates and helped bring inflation down from double digits to a more modest 4%, thus ending the period known as the Great Inflation. If you were around at the time, then you will recall that the Fed action also resulted in the worst recession in decades. In retrospect, many economists agree that the reduction in inflation or disinflation that resulted was worth the cost of recession. From the 1980s onward, inflation remained relatively low and stable and ushered in an economic era known as the Great Moderation.

Disinflation is beneficial to an economy for several reasons. Disinflation reduces pressure to increase wages, as prices are more stable. Disinflation also results in lower, more stable interest rates, which makes capital investment less costly and easier to plan. Arguably the most important outcome of disinflation is that producers' and consumers' inflationary expectations are lowered, which results in a profoundly more stable economic environment.

The Role of Expectations

One of the interesting features of economics is the possibility for self-fulfilling prophecies. In the realm of inflation, fear or hubris is often realized with changes in the rate of inflation. The fear of inflation or the general expectation that inflation will occur is often enough to spark an inflationary period. Consumers fearful of inflation will spend more and save less, which results in demand-pull inflation. The resulting demand-pull inflation reinforces the expectation of future inflation, and wage earners demand higher nominal wages to offset the effects. This of course results in costpush inflation. If policymakers fail to manage the expectations of inflation, a higher expected inflation rate embeds itself into the economic consciousness. With this higher expected rate of inflation, the economy is not able to produce as much output as it otherwise would and faces higher prices than it otherwise should.

Central bank authorities try to manage not only actual inflation but, more importantly, the expectation of future inflation. Because the fear of inflation is often enough to create it, policymakers are in the business of acting as a psychologist to the economy.

FACT

Managing inflation expectations means that the Fed chairman must be very careful about what he says. Former chairman Alan Greenspan was famous for his cryptic quotes, often referred to as Fed speak. According to Greenspan, "I guess I should warn you, if I turn out to be particularly clear, you've probably misunderstood what I've said."

It is not enough to talk the talk when it comes to managing expectations, however; the Fed must walk the walk. If you have ever dealt with children, you know that words without action are meaningless. A parent may respond to a teenager's rude behavior by threatening, "If you don't stop behaving this way, I will confiscate your cell phone for the weekend." If the rude behavior continues and the parent doesn't act on the threat, the parent's credibility is undermined. A parent who consistently follows through, however, is much more credible. Likewise, the Federal Reserve bolsters its credibility when it raises interest rates in response to inflation fears, but it loses credibility when it fails to respond forcibly to the possibility of inflation. The Fed's credibility as an inflation fighter was greatly reinforced by Paul Volcker's chairmanship because he said what he meant and meant what he said.

Deflation

If inflation is bad, disinflation is better, then deflation must be best, right? Wrong. **Deflation** occurs when the average price level is declining and money's purchasing power is increasing. What could be wrong with that? The problem with deflation is that it creates a perverse set of incentives in the economy. If prices are steadily declining, then consumers delay their purchase of durable goods as the deals just keep getting better as time passes. If this behavior continues, manufacturing grinds to a halt and widespread unemployment results. The unemployment would then reinforce the deflation, as fewer and fewer consumers would be willing and able to purchase goods and services. Producers respond similarly to deflation by delaying investment and compounding the effects of the delayed consumption.

Deflation poses a policy dilemma for central banks that primarily use interest rates to influence economic activity. In response to increased inflation, central banks raise interest rates to reduce the flow of credit and cool inflation pressure. There is no upper limit to how high an interest rate can go, but the opposite is not true. Given deflation, central banks will lower interest rates to encourage investment and consumption. If the lower interest rates do not have the desired effect, central banks will continue to lower until they hit what economists refer to as the **zero bound**. Once interest rates are at zero, they cannot go lower.

QUESTION

What is a good rate of inflation?

The magic number seems to be 2%. At a 2% annual rate of inflation, prices are relatively stable and slow growing. A 2% inflation rate results in prices doubling about every 36 years. A 2% rate is also high enough that should policymakers make mistakes, there is some cushion before deflation is a problem.

John Maynard Keynes referred to this weakness in monetary policy as **liquidity trap**. If consumers and investors will not borrow at 0%

interest, then you are out of options. The solution for deflation is to create inflation. Milton Friedman suggested that in economies with an inconvertible fiat money standard, deflation should never be a problem. All the monetary authorities would need to do is print money, or in the case of economies with independent central banks, monetize the debt, and the deflation would end. It's been said that this policy is like a government ending deflation by dropping cash from helicopters over the landscape. This policy solution was reiterated in 2002 by Benjamin Bernanke, which earned the Fed chairman the nickname "Helicopter Ben."

CHAPTER 16: Putting It All Together: Macroeconomic Equilibrium

Once you understand the concepts of supply and demand, GDP, unemployment, and inflation, you have a tool kit for understanding the economic fluctuations that occur. The aggregate demand and aggregate supply model will allow you to analyze the entire economy. You'll even be able to predict what might happen given certain events. If you are not careful, you might end up sounding like an economist the next time the Fed raises interest rates.

Aggregate Demand

Recall that demand is the willingness and ability of consumers to purchase a good or service at various prices in a specific period of time. **Aggregate demand** (AD) is a similar concept, but has some important distinctions. AD is the demand for all final domestic production in a country. Instead of just households, AD comes from all sectors of the economy. Furthermore, AD relates the price level to the amount of real GDP instead of price to quantity.

The relationship between the price-level and the amount of real GDP is inverse. The higher the price level, the less real GDP is demanded, and the lower the price level, the more real GDP is demanded. This is true because as the price level rises, money and other financial assets lose purchasing power. Fewer people demand our exports, and corresponding higher interest rates discourage investment and consumption. As the price level decreases, purchasing power increases, exports become more affordable to foreigners, and the corresponding lower interest rates encourage investment and consumption.

FACT

The next time you are watching the news and an expert mentions demand, chances are that she is talking about aggregate demand. The more you learn about economics, the more often you seem to hear about it.

Changes in AD occur when consumption, private investment, government spending, or net exports change independent of changes in the price level. For example, if the general mood of the country improves and consumers and businesses are feeling more confident, they will consume and invest more, regardless of the price level. This increase in consumption and investment increases AD. Likewise, increases in government spending or net exports also tend to increase AD. Reductions in any of the spending components of GDP will tend to suppress AD. If government raises the average tax rates on income, households' disposable income is reduced, and they consume less, which reduces AD.

Aggregate Supply

Supply is the willingness and ability of producers to generate the output of some good or service at various prices in a specific time period. **Aggregate supply** is a much broader concept than supply because it is inclusive of all domestic production, not just a singular good or service. Like an individual firm, an economy has a production function that relates the amount of labor employed with the amount of output or real GDP that the economy can produce with some fixed level of capital. In the short run, the amount of real GDP supplied is directly related to the price level. However, in the long run, the amount of real GDP producers collectively supply is independent of the price level.

The Short Run

Why do firms respond in the short run to price-level increases by producing more output and vice versa? Before answering this question, it's important to recall what is meant by short run in the macroeconomic sense. The short run is the period of time in which input prices (primarily nominal wages) do not adjust to price-level changes. If the economy experiences unexpected inflation, the short run is the period in which money wages remain fixed before finally adjusting to the inflation. During this period, firms realize higher profits as their output earns ever-higher prices while they maintain the same wage payments to their workers. Firms respond to the higher profits by increasing their collective output, or real GDP. In response to decreases in the price level, firms reduce output as they experience losses. This relationship is called the **short-run aggregate supply** (SRAS).

The Long Run

In the long run, the price level is irrelevant to the level of real GDP firms are willing to produce. The long run is the period of time in which input prices adjust to changes in the price level. Unlike the short run, where increases in the price level induce more output, in the long run firms do not realize higher profits and thus have no incentive to increase output. Why? Input prices match the increases in the price level. Therefore, in the long run, firms' input prices (wages) increase at the same rate as general price inflation and in real terms are constant. The independence of real GDP from the price level is referred to as **long-run aggregate supply** (LRAS).

Changes in SRAS

SRAS is affected by changes in per-unit production cost. As per-unit production costs fall, the economy is able to produce more real GDP at every price level, and as unit costs rise, the economy's ability to generate real GDP is reduced. In true economic style, per-unit production costs themselves are subject to influence by productivity, regulation, taxes, subsidies, and inflationary expectations. **Productivity** is output per worker and as it increases, per-unit costs fall. Decreases in productivity lead to higher per-unit production costs. Regulations place compliance costs on business and act to reduce SRAS. For example, to reduce sulfur dioxide emissions, factories must pay for smokestack scrubbers, which means that money cannot be used to increase output. Taxes on producers directly reduce their capacity to produce while subsidies increase their productive capacity. Finally, inflationary expectations influence the unit costs of production and therefore SRAS. As inflationary expectations increase, workers demand higher wages, lenders demand higher interest rates, and commodity prices increase as a result of speculation. The outcome is for SRAS to be reduced by inflationary expectations. Decreases in inflationary expectations have an opposite effect and serve to increase SRAS.

Changes in LRAS

LRAS is directly influenced by the availability of the factors of production. If land, labor, capital, and entrepreneurship increase, then LRAS increases. Decreases in the availability of these resources reduce the LRAS. Increases in LRAS are characterized as economic growth. Decreases in LRAS represent a long-term economic decline. The medieval black plague that wiped out a third of the European population is an example of an event that reduces LRAS. The invention of the steam engine exemplifies the type of technology that expands the LRAS.

Macroeconomic Equilibrium

Macroeconomic equilibrium occurs when the real GDP that is demanded by the different economic sectors equals the real GDP that producers supply. **Short-run equilibriums** occur when AD equals SRAS, and **long-run equilibriums** occur when AD equals LRAS. Changes in macroeconomic equilibrium occur when there are changes in AD, SRAS, or LRAS.

Increases in AD relative to SRAS result in both increased price level and increased real GDP in the short run, but just increased price level in the long run. As consumers, businesses, government, and the foreign sector demand more scarce output, firms respond to the increased price level by increasing output. In the long run, wages adjust to the increased price level, and GDP returns to its long-run potential at a higher price level. Demand-pull inflation results from increases in AD. Decreases in AD result in the opposite. As AD decreases relative to SRAS, both real GDP and price level fall. In the long run, wages and other input prices adjust to the lower price level and the economy returns to its long-run potential GDP at a lower price level than when the process began.

QUESTION

What happens to unemployment as aggregate demand changes?

Increases in AD lead to increases in real GDP. The increase in real GDP creates more demand for labor and reduces the unemployment rate. The reduced unemployment causes an increase in the price level.

Change in SRAS relative to AD also leads to changes in real GDP and price level. Unlike AD changes, which lead to GDP and price level moving in the same direction, SRAS changes result in GDP and price level moving opposite from each other. An increase in SRAS relative to AD will lead to a higher real GDP at a lower price level because as production costs fall, firms are more willing to produce more output at each and every price level. Decreases in SRAS relative to AD lead to the economic condition previously described as stagflation. Stagflation occurs when GDP decreases are combined with increases in the price level. When the costs of production rise, firms produce less output at each and every price level.

The Classical View

Prior to the Great Depression, orthodox economic thought could be described as classical. Today, **classical** refers not only to those economists with pre-Depression notions of the economy, but also can be used to describe a much broader group of economists who favor market-based solutions to economic problems. The classical camp espouses what is best described as a **laissez faire** philosophy.

The classical view of the economy is one that emphasizes the inherent stability of aggregate demand and aggregate supply. Efficient markets are able to quickly and effectively reach equilibrium conditions, so periods of extended unemployment are not possible. When consumers stop spending, they are saving instead. This increased saving reduces the real interest rate and spurs investment in capital, so any decreases in consumption are offset by increases in investment. This leads to the conclusion that AD is stable. If shocks do occur to the economy, flexible wages and prices allow the economy to quickly adjust to changes in the price level as rational economic actors take into account all available information when making decisions. For example, workers will accept lower wages in response to deflation and demand higher wages in response to inflation. This quick response implies that the economy tends to remain at its long-run equilibrium of full employment. Government interference is not warranted given this assumption, and as a result, laissez faire is the best policy.

ESSENTIAL

One of the assumptions at the heart of classical economic thought is **Say's law**. French economist Jean Baptiste Say believed that supply creates its own demand, and as a result, surpluses and gluts could not be sustained in a market economy.

The classical response to economic recession is to do nothing. A decrease in AD leads to lower GDP and a lower price level. The resulting high unemployment puts downward pressure on wages as workers will willingly go back to work for less money. These lower wages encourage firms to increase SRAS, and the economy returns to equilibrium at full employment. No government action is necessary as market forces are working to bring the economy back to full employment.

The classical response to inflation is also to do nothing. Increased AD leads to a higher GDP and higher price level. As the unemployment rate falls below the natural rate, considerable upward wage pressure results. With few unemployed workers available, firms compete with each other for already employed workers; this means offering higher wages to entice them to leave their current jobs. This intense competition for workers and other resources increases the costs of production for businesses. They eventually reduce production, and the economy returns to full employment at a higher but stable price level. Once again, no government intervention is necessary because market forces return the economy to its long-run full employment equilibrium.

The Keynesian View and Fiscal Policy

Britain's John Maynard Keynes was a classically trained economist who eventually came to the conclusion that the classical assumptions did not describe the reality of his day. During the Depression, Keynes wrote *The General Theory of Employment, Interest, and Money*. In it he challenged the prevailing assumptions and concluded that government intervention was warranted in the case of the Depression. He observed that saving did not instantly translate into new capital investment. Savings gluts could and did occur, and this implied that AD was inherently unstable, as decreases in consumption were not offset by increases in investment.

FACT

Keynes is a polarizing figure in economics. His ideas challenged the status quo, and he is seen by many as the enemy of free-market economics. The writings of Friedrich von Hayek and Ludwig von Mises of the Austrian school of economic thought are often quoted today as a counter to Keynes's arguments.

He also observed that wages and other input prices were not downwardly flexible. Workers did not readily accept pay decreases, nor did employers offer them. As a result, periods of high unemployment could persist as market forces did not function to bring the economy to full employment. The implication of these observations was that government intervention was necessary in the case of high unemployment.

Given a recession, the Keynesian response is to increase government spending and to reduce income taxes in order to spur aggregate demand and return the economy to full employment. This means that government must be willing to run deficits in order to carry out the policy. On the bright side, Keynes shows that returning the economy to full employment can be done relatively cheaply because of the multiplier effect. If real GDP is $14 trillion but potential real GDP is $15 trillion, government does not have to spend $1 trillion to close the recessionary gap but instead only a fraction of that because of the multiplier effect.

QUESTION

How do you calculate the spending or tax multiplier for an economy?

The spending multiplier can be calculated by dividing one by the marginal propensity to save. The tax multiplier can be calculated by dividing the marginal propensity to consume by the marginal propensity to save.

Here is how Keynes defined the multiplier effect. Keynes observed that individuals have a marginal propensity to consume and save. In other words, if you give people a dollar, they are inclined to spend some of it and save some of it. If government spent money on public works, the contractors and employees would then turn around and spend a portion of the resulting income and save the rest. This process would continue and lead to a multiplier effect throughout the economy.

For example, if Americans have a marginal propensity to consume 80% of their income, then an increase in government spending on infrastructure of $50 billion will result in $50 billion in government spending and then $40 billion of new consumer spending, followed by $32 billion and so on until eventually total spending equals $400 billion. How did that work? $50 billion dollars in government spending kicks off a continuing cycle of consumption and income. The higher the marginal propensity to consume, then the higher is the multiplier effect. Keynes also recognized that government spending yielded a larger multiplier effect than equal-sized tax cuts because people save a portion of their incomes. Thus, not all of the tax cuts' value is spent on consumption.

FACT

Keynesian economic policy was made the law of the land through the passage of the Full Employment Act of 1946. Signed by President Truman, the law requires government to pursue maximum employment, production, and purchasing power. The law also created the Council of Economic Advisors.

Keynes's observations and influence completely changed the study of economics. Today's policy framework is based on the ideas of Keynes and his followers. Although much of the modern debate is framed in terms of free-market capitalism versus socialism, Keynes was an advocate of capitalism and his approach is better characterized as a hybrid form of capitalism than socialism.

The Inflation-Unemployment Trade-Off

The influence of Keynesian thought grew from the 1930s until the 1970s. The research of New Zealand-born economist A. W. Phillips helped reinforce Keynes's influence on governments during this period. Phillips studied the relationship between wage inflation and the unemployment rate in Britain and concluded that periods of wage inflation were associated with periods of low unemployment. Periods of stagnating wages were associated with periods of high unemployment. American economists Paul Samuelson and Robert Solow adapted the **Phillips curve** for the U.S. economy. Using this model, they compared general price inflation (instead of wage inflation) to the unemployment rate. Many policymakers and economists reached the logical conclusion that Keynesian-style fiscal policies that stimulate AD could be used to sustain low unemployment at the cost of some known amount of inflation. The idea worked something like this: if policymakers wanted to reduce unemployment from 7% to 5%, the trade-off would be a known change in the inflation rate from 1% to 2%.

The experience of the 1970s caused some serious doubts about the legitimacy of the Phillips curve. Remember stagflation? During the 1970s, both inflation and unemployment simultaneously increased. These results did not align with the predictions of the Phillips curve, which implied the two were trade-offs. Milton Friedman and Edmund Phelps viewed this real-world data as a disproof of the Phillips curve and, more importantly, of the validity of Keynesian economics. Friedman and Phelps introduced the **natural rate hypothesis**, which concluded the rate of unemployment is independent of inflation in the long run. Efforts by government to reduce unemployment by creating temporary inflation would be ineffective. Workers would try to keep their real wages from falling by demanding higher nominal wages in line with inflation.

FACT

A.W. Phillips was something of an engineer. In addition to his economic theories, he also invented a device called the MONIAC, which could be used to teach students how the macro economy works. The MONIAC was a mechanical device that pumped fluid through a series of tubes demonstrating the flows of money in the economy. The MONIAC could be set up to operate under both classical and Keynesian assumptions.

The new consensus on the Phillips curve is that there are two types of curves. There is a **short-run Phillips curve** that implies a trade-off between inflation and unemployment, and there is a **long-run Phillips curve** that exists at an economy's natural rate of unemployment. Sampling a few years of inflation and unemployment data may suggest an inverse relationship between the two, but including all available data reveals no relationship between the two variables.

Economists chalk up changes in the short-run Phillips curves around the long-run Phillips curve to being the result of changing inflation expectations. Phillips's original observations about the British economy were about a time period where expected inflation was stable. The breakdown of the adjustable peg exchange rate regime in 1971 effectively ended any type of gold standard and introduced a period of uncertainty about inflation. The resulting increases in inflation expectations help to explain the increased inflation and unemployment that occurred. Short-run Phillips curves exist during periods of stable inflation expectations. When inflation expectations change, the short-run Phillips curve relationship breaks down, and either simultaneous increases or decreases in inflation and unemployment can occur. Once a new expected inflation rate embeds itself into the economy, a new short-run Phillips curve emerges.

ESSENTIAL

An assumption of Keynesian economics is that people suffer from money illusion. They prefer earning $100 per hour and paying $10 per gallon of gasoline to earning $10 per hour and paying $1 per gallon of gasoline. Even though real purchasing power is the same in each example, people have a strong preference for higher nominal wages.

Alan Greenspan attributes the reduction in inflation expectations as the reason for the low inflation and unemployment that occurred during his chairmanship at the Fed. According to Greenspan, productivity gains from globalization subdued inflation fears for much of his tenure. In his book, *The Age of Turbulence*, he argues that Ben Bernanke is in the unenviable position of holding office during a time where inflationary expectations are on the rise. Bernanke's policy decisions occur along short-run Phillips curves that offer higher rates of inflation and unemployment than what Greenspan faced. Taking into account the influence of inflationary expectations, Keynesian policies will only work as long as inflation expectations remain stable. If the Keynesian AD policies create higher inflation expectations, they will be thwarted as attempts to stimulate AD and reduce unemployment will only create more inflation.

CHAPTER 17: The Federal Reserve and Monetary Policy

America has a long and storied love-hate relationship with its banking system. The most vilified institution is the nation's central bank, the Federal Reserve, or just the Fed. The Fed is not America's first central bank or even its second. Regardless of your feelings towards it or the history behind it, the Fed is at the center of the American economy and deserves your careful consideration.

History of Banking Revisited

At the end of the American Revolution, the United States was saddled with significant war debt. In order to handle the debt and to create a unified currency, Alexander Hamilton proposed the creation of a central bank for the young nation. The First Bank of the United States, headquartered in Philadelphia, was modeled after the Bank of England. It served as the nation's central bank from its charter in 1791 until 1811 when the charter was allowed to expire.

FACT

The United States has a long history of central banking, but Sweden wins the prize for the oldest central bank. The Sveriges Riksbank has been in operation since 1668. Sweden's central bank is important also because it sponsors the Nobel Prize in economics.

In the nineteenth century, America went through a series of economic panics that led to the creation of another central bank. This bank eventually became a political target and failed to bring economic stability to the country.

The Second Bank of the United States was given a twenty-year charter in 1816 to help the United States economy recover from the economic effects of the War of 1812. The Second Bank established a uniform currency and acted as a depository for the Treasury's accounts. The Second Bank of the United States was believed by many to be corrupt, and political pressure from President Andrew Jackson sealed its fate. In 1833, three years before its charter expired, Jackson had his Treasury secretary withdraw the United States government's deposits from the bank and placed in state-chartered banks. This effectively killed the bank, and by 1841 the Second Bank of the United States was bankrupt.

During and after the Civil War, the United States created more national banks and reintroduced a single uniform currency. These national banks were instrumental in allowing the government to borrow by issuing bonds. Unlike the First and Second Banks of the United States, however, these national banks were decentralized.

The Bank Panic of 1907 was the impetus for creating the Federal Reserve System. A failed attempt by a Montana investor to corner the copper market led to a loss of confidence in the financial institutions associated with him and his brother. The contagion spread to unassociated institutions, and within a few days an all-out run on the New York financial system was under way. Fortunately, J. P. Morgan and Theodore Roosevelt's Treasury secretary brought calm to the situation by injecting cash into the banking system and eventually bringing an end to the bank runs. The panic of 1907 showed that America needed a central bank to act as a lender of last resort to ensure liquidity in the banking system.

The Federal Reserve System

The 1913 passage of the Federal Reserve Act, signed into law by Woodrow Wilson, created the **Federal Reserve System**. Unlike previous attempts at central banking, the Fed drew from the strengths of its predecessors. Instead of creating a single central bank, the Federal Reserve Act established a decentralized, public-private banking system. The Fed is not headquartered in a single location but has separate locations across the United States. The Federal Reserve is neither a purely governmental institution nor is it a purely private institution. The Fed has features of both.

QUESTION

Where does the Fed keep all of the money?

Most of the money on deposit with the Federal Reserve exists in electronic form, but each of the district banks has a significant vault with millions of dollars under heavy security. In the early days of the Fed, the teller window was protected from bank robbers by pill boxencased machine guns.

The Fed is the bank for the United States government. The treasury keeps its accounts with the U.S. government, and the government in turn writes checks from its accounts with the Fed. The taxes collected and the money borrowed through issuing government bonds are all deposited in the U.S. Treasury's account with the Fed. Every time a taxpayer receives a refund, or a Social Security recipient receives their check, the checks are being drawn from the treasury's account with the Fed.

The Federal Reserve System's government arm, the **Board of Governors**, is headquartered in Washington D.C. The governors are appointed by the president of the United States and confirmed by the Senate for single, staggered fourteen-year terms. The board is supervised by the Fed chairman and vice chairman, who are also members of the board. The chairman and vice-chairman are appointed by the president and confirmed by the Senate for unlimited four-year terms. The Fed chairman is the face of the Federal Reserve System and is considered by many to be second in power only to the president of the United States when it comes to economic influence. The Board of Governors creates policy and regulations for the nation's banking system, sets reserve requirements, and approves changes in the discount rate.

In keeping with the United States' federal nature, the Fed is divided into twelve distinct geographic districts with headquarters for each district located in cities across the United States. Each district is an equal part of the Federal Reserve System. The district banks act as the bankers' banks and accept deposits from member banks. The district banks also perform a regulatory role in their district by monitoring the member banks and enforcing regulations within their respective districts. The district banks serve a vital role in processing paper check and electronic payments for the banking system. Finally, the district banks issue the currency to the banking system that they acquire from the U.S. Department of Treasury.

The FOMC

The **Federal Open Market Committee** (FOMC) is the chief architect of the nation's monetary policy. The twelve voting members of the committee are made up of the Fed chairman, the Board of

Governors, the Federal Reserve Bank of New York's president, and four other district bank presidents who serve on a rotating basis, although all the district bank presidents are present at the committee meeting. The FOMC meets eight times a year, or about once every six weeks to review economic performance and decide the course of monetary policy by targeting the fed funds rate. FOMC meetings are closely monitored by the press and financial markets. Members of the media and investors carefully analyze the FOMC's press releases, looking for clues as to what might be the future direction of policy.

The structure of the FOMC makes it an effective decision-making body. The presence of the Board of Governors ensures that the needs of the country as a whole are being met, while the presence of the bank presidents ensures that regional concerns are considered. The governors are also representative of the public sector, while the presidents are representative of the private sector. Each district prepares a report on the economic conditions in the district, called a **Beige Book report**. In coming to a decision on monetary policy, many voices are heard and the perspectives of the different constituencies are represented.

FACT

Investors attempting to profit by speculating on the Fed's interest rate policies would study the size of Alan Greenspan's briefcase during his tenure as Fed chairman. They theorized that if the briefcase was big and heavy, he was carrying documentation to support his argument for changing interest rates. If the briefcase was light, then interest rates would probably remain unchanged.

Monetary Policy

The goal of **monetary policy** is to promote price stability, full employment, and economic growth. In order to achieve these goals, the Fed conducts two primary forms of policy. The Fed conducts **contractionary**, or tight, monetary policy to restore price stability. To restore or maintain full employment and foster long-run economic growth, the Fed engages in **expansionary**, or easy, monetary policy. The Fed achieves these goals by using several tools. The Board of Governors sets the reserve requirement for member banks, the district banks influence the discount rate with board approval, and the FOMC directs open market operations conducted by the New York Fed.

The Fed, a monopolist over the money supply, is in a unique position to influence aggregate demand in the economy. Fed policy affects excess reserves in the banking system, which directly influences the money supply, which, in turn, changes interest rates. These changes in interest rates lead to changes in aggregate demand via consumption, investment, and net exports. The resulting change in AD affects GDP, inflation, and unemployment.

To understand how monetary policy functions, consider the following analogy. Assume that the economy is a car operated by a nearsighted teenager with a learner's permit. When the teenager hits the gas, the economy grows. When he floors the gas pedal, the economy experiences inflation. If the teenager is timid and refuses to touch the gas pedal, he causes a recession. In this analogy, the Federal Reserve is represented by a kindly old driving instructor with severe nearsightedness but an extensive driving record. Like many drivers' education cars, this one comes equipped with a brake pedal for the instructor. How can the instructor influence the driver? She can apply pressure to the brakes and talk to the driver.

The Federal Reserve influences the economy by applying pressure to interest rates and talking to the public. In the case of economic growth, the Fed can warn against inflation by ever so slightly raising interest rates whenever the economy seems to be getting a little too fast. During periods of inflation, the Fed can mash down on the interest rate brake pedal, give a stern warning to the driver, and bring the economy back under control. In the case of recession, the most the Fed can do is let off the brake pedal, or lower interest rates and encourage the driver to hit the gas. The Fed's power is asymmetrical; it is capable of stopping inflation, but only capable of encouraging full employment.

Reserve Requirement

A bank's **reserve requirement** is the percentage of checking account balances that the bank is not able to lend against. If the reserve ratio is 10% and checking account balances total $2 billion, then the bank may lend excess reserves of $1.8 billion but must hold required reserves of $200 million. Raising the reserve requirement would effectively limit inflation, while lowering the reserve requirement would reduce interest rates and encourage investment and consumption. If the Fed were to raise the requirement, banks would have fewer excess reserves from which they could lend. This would reduce the money supply and result in higher interest rates, discour-aging capital investment and durable goods consumption. This decrease in investment and consumption lowers AD and leads to less real GDP, higher unemployment, and a reduction in inflation. Lowering the reserve requirement would have the exact opposite effect.

The problem with changing the reserve requirement is that it is easy to lower but much more difficult to raise. If the Fed lowers it, banks are able to lend more of their reserves and no real problem is created. However, raising the reserve requirement during periods of inflation would be nearly impossible for banks, as they probably have no excess reserves not already lent. The increase in the reserve requirement would precipitate an immediate liquidity crisis in the banking sector. Banks would call in loans and do whatever they can to meet the new higher requirement. The best time to raise the reserve requirement is when banks have available excess reserves, such as during a recession. However, raising the reserve requirement is contractionary, and so raising it during a recession is harmful to the goal of maintaining full employment.

FACT

Sometimes banks will keep excess reserves in addition to required reserves. Banks maintain contractual clearing balances with their Federal Reserve district bank in order to meet day-to-day demands for cash.

Discount Rate

District banks, with Board of Governors approval, are able to raise or lower the discount rate in order to influence the flow of credit. The **discount rate** is the interest rate that member banks pay the Fed to borrow money overnight, usually when they are in financial distress. Raising the discount rate discourages borrowing, but lowering the discount rate encourages it. If the Fed wants to reduce inflation, it raises the discount rate. When the Fed does this, the following chain of events is set into motion. The Fed announces an increase in the discount rate, banks are discouraged from borrowing, excess reserves are less likely to be lent, the money supply does not grow, interest rates rise, consumption, investment and net exports fall, AD decreases, GDP falls, unemployment rises, and inflation falls. Lowering the discount rate would have the exact opposite effect.

The problem with the discount rate is that banks are reluctant to borrow directly from the Fed. The reluctance stems from the fact that other forms of borrowing are available at lower rates, so going to the Fed's discount window is a public admission that something is wrong with the borrowing bank. A bank in good financial position will usually borrow to cover its short-term needs in the interbank lending market, paying the fed funds rate or the overnight LIBOR. The principal decision-makers at a bank do not want to scare away investors or current shareholders by borrowing from the Fed. Going to the discount window reduces the value of the bank's stock and causes the bank's board of directors to possibly look for new leadership.

The discount rate is useful as a signal of future interest rate policy. In February 2010, the Fed raised the discount rate while maintaining the current fed funds rate. The increase in the discount rate signaled to investors and the economy that the Fed was keeping an eye on inflation, which helps to alleviate inflationary expectations while also signaling that increases in the fed funds rate would be forthcoming. This signaling function is important, as financial markets do not like surprises.

Open Market Operations

The primary way that the Fed enacts monetary policy is through the process known as **open market operations** (OMO). OMO is the buying and selling of U.S. treasury securities (treasuries) between the Federal Reserve Bank of New York's open market desk and a select group of eighteen primary security dealers. The dealers include most of the world's major banks and securities broker-dealers. They are expected to participate as counterparties to the Fed's open market operations and share market information with the Fed. OMO have the effect of either increasing excess reserves in the banking system when the Fed buys treasuries from the primary security dealers, or reducing excess reserves in the banking system when the Fed sells treasuries to the primary dealers.

If the FOMC decides to target a lower fed funds rate, the New York Fed's open market desk is directed to purchase treasuries from the primary security dealers. The dealers then deposit the proceeds from the sale into their bank accounts, which then permanently increases excess reserves in the banking system. The open market sale of treasuries by the New York Fed to primary security dealers results in a permanent reduction in excess reserves. Fine-tuning to increase reserve balances in the banking system is carried out when the Fed engages in **repurchase agreements,** or repos, with the primary security dealers. In a repo transaction, the Fed lends money to the dealers, sometimes overnight, in exchange for a treasury security or other high-quality financial asset. This has the effect of temporarily increasing the amount of available excess reserves in the banking system. If the Fed wants to temporarily reduce the amount of excess reserves in the system, they engage in **reverse repurchase agreements**, or reverse repos, with the dealers. In a reverse repo, the Fed borrows cash from the dealers in exchange for treasury securities on a short-term basis. This temporarily decreases the available excess reserves from the system.

QUESTION

Who thought of using open market operations to influence the economy?

Open market operations were an accidental discovery made in the 1920s. During an economic slowdown, the district banks were conducting so little business that their survival was in jeopardy. To preserve the system, they began buying up treasury securities in order to earn interest income. This led to expanded excess reserves in the banking system. By the 1930s, open market operations were a formal part of Fed monetary policy.

When Policies Collide: Fiscal/Monetary Mixer

Federal Reserve policy does not exist in a political vacuum. Instead, monetary policy functions alongside government's **fiscal policies**. At times the two are at odds, but most of the time monetary policy is used to accommodate fiscal policy. For this to work, the Fed chairman, the president, the Treasury secretary, and key members of Congress communicate to create a coherent policy that addresses the fundamental goals of price stability, full employment, and economic growth.

During periods of recession, **expansionary fiscal policy** is reinforced with **expansionary monetary policy**. As government spending increases and taxes decrease, upward pressure is placed on interest rates, which may "crowd-out" private investment and consumption and reduce net exports. Expansionary monetary policy offsets the upside pressure on interest rates by expanding the money supply and lowering short-term interest rates. This allows for fiscal policy to be more effective.

ESSENTIAL

Monetary policy is conducted by the Fed, and fiscal policy is conducted by the government. The coordination of the two requires communication. The chairman of the Fed gives a report to Congress twice a year.

Inflationary periods are more problematic for presidents and lawmakers. The **contractionary fiscal policy** prescription calls for reduced spending and increased taxes, which are politically unpopular. **Contractionary monetary policy** is effective at stopping inflation, and the Fed's insulation from political pressure makes it perfectly suited for the task. Former Fed chairman Paul Volcker was credited with whipping inflation when government was unable to do so. Problems arise when government maintains an expansionary policy while the Fed is engaging in contractionary policy. President George H. Bush credits some of his defeat to President Bill Clinton on the Fed's contractionary policy stance during the election of 1992.

Monetary Policy in the Short Run and the Long Run

Monetary policy has different effects in the short run and the long run. Expansionary monetary policies designed to reduce short-term interest rates and spur full employment and economic growth eventually lead to higher interest rates as they induce inflation. This means that monetary policy must be carefully applied. If the Fed introduces a monetary stimulus, they must also plan to remove the stimulus in order to prevent future inflation. The problem for policymakers is in timing the policy. Early removal of monetary stimulus might result in a protracted recession, but maintaining low interest rates for too long will almost certainly lead to higher inflation.

ESSENTIAL

> Central bank independence is key to controlling inflation. In countries with independent central banks, the temptation of government to pay its debts by printing currency is severely limited. If Congress ran the Fed, then the temptation to print up currency instead of raising taxes might prove too big a temptation, and inflation would ensue.

Too much reliance on monetary policy to reduce inflation can also lead to problems. Over the business cycle, if government uses expansionary fiscal policy to offset recessions and then relies on the Fed to contain periods of inflation, interest rates will ratchet up and long-term economic growth will be stymied. At some point government must reign in its spending and/or raise taxes to keep long-term interest rates from rising too high. Some observers think that the Fed's goals of price stability and full employment are mutually exclusive and impossible to simultaneously achieve. Those within the Fed who have a similar view tend to be labeled as inflation hawks for their insistence that price stability be the overarching goal of the Fed.

CHAPTER 18: Voodoo Economics

In the 1986 film *Ferris Bueller's Day Off*, a dazed class sits and listens to their economics teacher, played by Ben Stein, drone on about the Laffer curve and voodoo economics. He's referring to the much-maligned theory of supply-side economics, which had its heyday during the Reagan administration. If you pay careful attention, you will discover that supply-side arguments are still heard today.

Stagflation

The concurrent high unemployment and inflation of the 1970s was a painful period in American economic history. Stagflation came as a shock to many politicians and economists alike. By the 1970s, Keynesian economic thought was embedded in the minds of most policymakers. Although Milton Friedman and Edmund Phelps had publicly refuted the idea that there is stable trade-off between inflation and unemployment, policymakers were not quite willing to let go of the belief.

eV FACT

Arthur Okun developed a simple index of economic hardship called the **misery index**. The misery index is the sum of the unemployment rate and the inflation rate.

Given normal conditions, the misery index is usually around 7. During the Carter administration, the misery index steadily increased because of stagflation and reached an all-time high of 20.76.

The Keynesian economic framework so permeated policy decisions by the 1970s that data challenging the effectiveness of Keynesian economics created cognitive dissonance for many. When unemployment is high, government should spend more, and when inflation becomes a problem, the Fed should tighten. What stagflation presented was an intractable problem for many in positions of power. Spending to alleviate unemployment would only make inflation worse. Fed tightening would reduce employment. The policy options of those influenced by Keynes focused on either increasing or decreasing aggregate demand. What to do? The answer offered by some was to focus on the supply side of the aggregate supply and aggregate demand model.

Thinking Outside the Box

In the 1930s, John Maynard Keynes challenged the conventional wisdom and introduced the world to his style of economics. Instead of acting in a hands-off manner, government could and should step in to the economy as a stabilizing force. What the economy would not fix on its own, the government would now manage. The ideas of Keynes must have been appealing to politicians who wanted to be seen as doing something about the economic problems that faced the average citizen. From a political standpoint, classical economics is suicidal in the face of economic recession. "Do nothing, and everything will work out like it has before," is probably not what most voters want to hear. Keynesian economics gave philosophical credibility to politicians who wanted to do more than just watch the economy recover. Keynesian economics became a reason for politicians to act. And act they did.

ESSENTIAL

Although it might appear that all things Keynes are associated with Democrats and supply-side economics with Republicans, this is not the case. Much of the criticism of supply-side economic theory came in the 1980 Republican primary from Reagan's chief opponent, George H. W. Bush, who referred to Reagan's economic plan as **voodoo economics**.

By the 1970s, government was entrenched in the economy. Regulations, taxes, price controls, and subsidies had steadily increased from the 1930s to the 1970s. Incrementally, and at other times in bursts, the government intervened in the economic activity of the nation. Notable examples include FDR's New Deal, JFK's tax cut, LBJ's War on Poverty, and the wage and price controls under Nixon. Keynesian economics had gone from outside-of-the-box thinking to the dominant economic philosophy in a relatively short period of time.

The shoe was now on the other foot. Free-market capitalism, the once dominant economic philosophy, was on the outside. It needed to be reinvented because doing nothing in the face of economic hardship has not been nor will ever be a viable political solution. The political answer came in the form of supply-side economics. Politicians could now advocate undoing that which other politicians had previously done. Where previous administrations had regulated, raised taxes, managed prices, and granted subsidies, the Reagan administration would deregulate, cut taxes, set the markets free, and reduce subsidies.

Supply-Side Economics

Stagflation was fundamentally a supply problem, which is why a demand-side solution would not work. Classical economics' laissez faire approach enjoyed something of a renaissance after Ronald Reagan's election. In the classical view of the economy, flexible and efficient markets ensure that the economy will maintain full employment. When recessions or periods of inflation set in, flexible input prices cause aggregate supply to increase or decrease and bring the economy back to full employment without government intervention. In the years from the 1930s to the 1970s, there was a decrease in the flexibility and efficiency of the labor market. The **supply-side** argument was that government had gummed up the works and the markets needed to be de-Keynesed.

Deregulation, which began under President Carter, picked up steam under Ronald Reagan. The airlines, which in the past had been heavily regulated by government, were set free and forced to compete with each other. This resulted in far more flights at cheaper prices. It also meant that the least efficient airlines were driven out of business.

Labor unions saw a decline in power under the Reagan administration. Foreign competition in the steel and car industry weakened the position of the unions. Probably the most powerful symbol of the loss in union power came at the hands of the newly elected president. The air traffic controllers union had lobbied for better pay and working conditions. In 1981, the union went on strike in violation of federal law. After being warned to return to work or be fired, over 11,000 air traffic controllers refused to return and were summarily fired by President Ronald Reagan. The message was clear.

ALERT

A debate is raging today between the Right and Left over the role of government in the economy. Those on the left with a Keynesian perspective are derided as "socialists," while those on the right who favor less government intervention in the markets are decried as "market fundamentalists."

While running for office, Reagan promoted the idea that tax cuts on high-income earners would enrich all Americans, as people had more incentive to spend and save. The spending and saving would lead not only to more consumption but also more capital investment. As investment increases, businesses expand their productive capacity, and this leads to higher employment. The resulting increase in capital also leads to greater productivity and, eventually, lower prices.

The power of tax cuts to stimulate aggregate demand was well known to Keynesian economists. But the supply-side spin was that tax cuts would not only stimulate aggregate demand but supply as well. At the same time that Reagan was proposing tax cuts, he also called for an increase in defense spending to counter the Soviet threat. Economists, politicians, and average people questioned the idea of simultaneously cutting taxes and increasing government spending. It appeared obvious that such a combination of policies would result in the federal government running large deficits as it spent more and taxed less.

Reagan, Cheney, and Laffer

In 1974, President Ford was considering a tax increase to offset rising inflation and budget deficits. Tax increases are politically unpopular, and for a president who had just pardoned the least popular politician in American history, Ford needed some new ideas. A now well-known Ford staffer by the name of Dick Cheney met with Arthur Laffer, an unorthodox economist, at a hotel bar in Washington D.C. What Laffer showed Cheney in 1974 was to become an influential argument in the 1980 election.

Laffer sketched a bell curve on a napkin that showed the relationship between tax rates and tax revenue collected by government. According to Laffer, as tax rates increase from 0% up, the government collects more tax revenue. However, if government sets a tax rate of 100%, then tax revenue will be $0.00 because at a 100% tax rate people have no incentive to work. Would you show up to work if the government took all of your paycheck every month? This implied that there is some point at which increasing tax rates reduces the amount of tax revenue that government collects.

Laffer, who became one of Reagan's economic advisors, argued that more tax revenue could be generated if tax rates were reduced because people would have more incentive to work, save, and invest. Economic growth would expand the tax base. Eventually, tax revenue would increase and the budget deficits would fail to materialize. The **Laffer curve** became a rationale that lent credibility to supply-side economics, just like the Phillips curve had been a rationale for supporting Keynesian demand-side policies.

Challenges to Supply-Side Economics

Sometimes reality has a way of ruining a great idea. The Laffer curve was used to defend tax cuts during a period of increased government spending. Reagan got his tax cuts and he got his increases in defense spending, but he also got huge deficits. Increased tax revenues failed to materialize. Instead, tax revenues drastically fell, and government budget deficits increased steadily during Reagan's administration. So is the Laffer curve logically flawed? No, the Laffer curve's logic is sound. The issue with the Laffer curve is that it is not very useful for making policy decisions. To be useful, it would help if policymakers were able to identify where they are on the Laffer curve. Unfortunately for Laffer and the budget, America was already on the "right" side of the curve. Instead of arguing about the existence of the Laffer curve, the real debate should have been about which side of the curve America was on.

Many economists attribute the economic growth of the 1980s not to the effectiveness of supply-side economics but to the reduction in expected inflation brought about by Fed Chairman Paul Volcker. Appointed by Jimmy Carter and later reappointed by Ronald Reagan, Paul Volcker had the fortitude to squeeze inflation from the economy by increasing the fed funds rate to an unprecedented 19%. This had the effect of both reducing actual and expected inflation, which allowed for growth to occur.

A criticism of supply-side economics is that it effectively redistributes income to the rich. Because the mantra of supply-side economics is "tax-cuts on income and capital gains," it stands to reason that the immediate beneficiary is going to be those with significant income. In the United States, the wealthiest 50% of households pay more than 95% of the income taxes and the wealthiest 5% pay 50% of the income taxes. If one group is getting taxed and the other is not, a tax cut will effectively increase the disposable income of taxpayers relative to nontaxpayers. If Jane is in the top 5% of incomes and Bob is in the lowest 5%, a taxcut raises Jane's income but has no effect on Bob because Bob was not paying taxes in the first place. This creates a condition where the rich get rich faster than the poor do. If that is not fodder for class warfare politics, then nothing is.

FACT

> The simultaneous tax cuts and spending increases that occurred under Reagan, although motivated by supply-side economics, are the Keynesian prescription for recession. From 1980 through 1982, the U.S. economy suffered a severe recession brought on by contractionary monetary policy to combat the high inflation of the era. Aggregate demand was the immediate beneficiary of the Reagan-era fiscal policy.

Although supply-side economics is not mentioned much anymore, its arguments and logic are still part of the Republican and Libertarian political platforms. Cutting taxes, creating incentives for people to save and invest, and a general distrust of government involvement in the economy are supply-side ideas that still resonate with many voters. Democrats tend to promote a more populist agenda. Tax cuts for the middle class with tax increases on the wealthy, increased regulation of business, and the use of transfer payments to redistribute income are all ideas advanced by the Democrats and associated for good or bad with the Keynesians.

During the 2008 Democratic primary debate between Barack Obama and Hilary Clinton, the candidates were debating their approach to stimulating the economy and their respective tax plans. Debate moderator Charlie Gibson asked Obama if he supported the

Bush tax cut on capital gains. Obama responded that he did not. Gibson then revealed a Laffer curve result. The capital gains tax cut had actually increased tax revenue to government. Obama, always a gifted debater, retorted that it was unfair that rich people paid a lower marginal rate on capital gains than workers do on their income. Many people believed that the logical conclusion of this debate is that the tax rate on capital gains is not set to raise revenue for government but to punish or reward those who save and invest. This goes to show that when it comes to economics and politics, politics wins.

A Complete Toolbox

Although supply-side economics as a field of study is derided by most mainstream economists, it has served to remind people that incentives matter. Policymakers must consider not only what voters want but also how their policies shape the incentives of consumers and producers. Whenever government creates a new mandate that seeks to regulate economic behavior, it must also be prepared to deal with the unintended consequences that occur as the new mandate alters the incentives of individuals and institutions.

Ignoring the supply side of the economy leads to a unilateral approach to policymaking that ultimately boxes government into two choices: increase aggregate demand or decrease aggregate demand. By recognizing the role of aggregate supply, policymakers can promote more policy solutions to achieve their ultimate economic goals. For example, recognizing that a tax cut on personal income has demand-side and supply-side effects allows policymakers to sell the option to their diverse constituencies.

ESSENTIAL

Some politicians have called for abolishing the income tax and replacing it with a national sales tax. Supporters of a sales tax argue that it would be easier to collect, and, more importantly, it would encourage people to save more and spend less. Opponents argue that sales taxes are regressive because they place a heavier burden on lower-income households than on upper-income households.

When it comes to taxation, two views of the economy predominate. One view is analogous to a pie of fixed size. The other view recognizes the economy as something more dynamic and capable of changing like a freaky space pie that can grow or shrink. In order to increase tax revenue in the fixed-pie universe, government raises tax rates on the bigger slices. In the other universe, the tax rate affects the rate of freaky space pie growth. Two options exist for raising tax revenue in the second universe. Government can raise the tax rate on the bigger slices and collect more revenue immediately, but because it reduces the rate of economic growth, government sacrifices future revenues. A counterintuitive approach to generating tax revenue is to tax at a lower rate, thus sacrificing immediate revenue in order to gain more revenue in the future as the low tax rate encourages expansion of the tax base.

The problem with analogies is that they often ignore important facts. An obvious one is that instead of inanimate slices of pie being taxed, in the real world individuals are capable of reacting to tax rates. What if the big slices of pie stood up and moved to a different plate in response to being taxed? High tax rates may ultimately chase off the most productive members of the economy. This is already happening within the United States as businesses uproot from high-tax states toward low-tax states. States like Michigan are losing out to states like Texas. Incentives must be addressed, and supply-side economics has helped to bring this to the forefront.

CHAPTER 19: Economic Growth

In Jared Diamond's study of human history, *Germs, Guns, and Steel*, he talks about the question that prompted him to study the course of human events. As an avid bird watcher, Diamond took many trips to New Guinea where he befriended a man named Yali. Yali asked Diamond why it was that the European descendants had so much while the people of New Guinea had so little. Diamond's fascinating account of the forces that shaped human history and the distribution of wealth is great reading. However, if Yali's question was answered by an economist, it would have only taken a chapter.

What Growth Means

Economic growth occurs when there is a sustained increase in a nation's real gross domestic product per person over time. In most years the United States real GDP grows at a rate of approximately 2%. This means that on average, the economy doubles in size every 36 years. At the same time, the population increases at a rate of 1%, which means that current real GDP per capita is almost three times greater than in 1960. Economic growth is not guaranteed. Indeed, there are years in which no or negative growth occurs. These periods are associated with recession.

Growth reveals itself in positive and negative ways. Economic growth leads to increases in living standards, nutrition, health care, longevity, and material abundance. The downside is that economic growth often results in environmental destruction and increased income inequality. Economic growth as a goal for society is hotly debated, and both sides offer wellreasoned rationales for their positions.

Why Grow?

Proponents of economic growth focus on the benefits it creates for society. The advances in food production, health care, longevity, and material abundance would not be possible without economic growth. A century ago, most Americans were involved in agricultural production and yet subsisted on far fewer calories than today's Americans, less than 2% of whom are farmers. The average life span has increased from 48 to 78 years in the same period because of the eradication of many diseases and advances in basic sanitation and health care. The quality and quantity of material goods has increased as well, allowing more Americans the things that only the wealthy could acquire in previous generations. The average workweek has decreased in the same period of time, allowing people more leisure. For most, economic growth has been a blessing.

As the economy grows and diversifies, more and more people are able to escape subsistence farming and pursue other areas of interest. This freedom for most everyone to pursue their interest and passion did not exist for most of recorded history. The creative explosion of production that has occurred over the last 150 years has yielded advances in all fields of human endeavor. Where life was nasty, brutal, and short for most, it is now relatively humane, peaceful, and long. If you have ever visited an old cemetery, you might have noticed the number of graves for young children. What was once a common occurrence is now a rare tragedy. The diseases that ravaged the population less than a century ago are for the most part eradicated. All of this is possible because of economic growth.

QUESTION

Can money actually buy happiness?

When economists compare GDP per capita with a country's overall level of happiness, an interesting trend occurs. As GDP per capita increases from $0 to $10,000, the level of happiness increases. This relationship breaks down after that. So to answer the question, the first $10,000 does buy happiness. After that, who knows?

Material abundance is a benefit of growth and is often seen by critics as the driving force behind growth policy. They may be right, but if growth also yields substantive benefits for humanity, an extra iPod or McMansion might be a small price to pay. Material abundance is a natural outflow of humans being set free from subsistence agriculture to think, invent, and create. And even for those who choose to return to the land and farm, that freedom exists as well.

FACT

The average number of children per family has steadily declined over the last 100 years. One reason for this is the reduction in infant mortality. In countries with high infant mortality rates, the average number of children per family is much higher.

As people have become more specialized and more productive, their value to society has increased as well. Consider the amount of time and resources now devoted to raising an American child. The average American child has over $250,000 invested in her human capital. Aid organizations understand that increasing an individual's value to society is important for developing a stable, productive society. Organizations that provide food assistance in the developing world have discovered that delivering families' food rations to the daughters in school increases the value of the daughter to the family, which reduces the incidence of childhood pregnancy and prostitution.

Human Capital

The most important element in economic growth is human capital. **Human capital** consists of the education, skills, and abilities possessed by an individual. Countries that invest heavily in human capital tend to have more economic growth than similarly endowed countries that do not. The United States is one of the world leaders in developing human capital. Compulsory primary and secondary education, mandatory vaccinations, and abundant nutrition have contributed to making America the most productive nation on earth. No nation spends more on educating and developing human capital than the United States.

Individual freedom and the ability to acquire private property are also essential elements in developing human capital. When individuals are free to choose their vocation and enjoy the benefits of private property, their productivity is higher than in places where individual freedom or private property is not valued. By way of comparison, the average German was far more productive in capitalist West Germany than in communist East Germany, and the average South Korean today is far more productive than the average North Korean, because economic freedom provides the incentive to produce more in order to have more.

To have human capital, you must have humans. For economies to develop and grow, it is also important that the population grow as well. Population growth must also occur alongside productivity growth. Larger populations are capable of producing more output as well as more innovation because the greater the population, the greater the number of productive resources. The more people a country has, the more probable entrepreneurs it has. And entrepreneurs are one of the drivers of economic growth.

The presence of immense natural resources can sometimes be a deterrent to developing human capital. The paradox of natural resource wealth is that governments are often so eager to exploit their natural resources for export that they neglect to invest in their population's human capital. At first glance, a comparison of Russia and Japan would lead most observers to believe that Russia is far wealthier than Japan. After all, it has the largest deposits of minerals and natural wealth in the world. Japan on the other hand has few natural resources. However, Japan has enjoyed much greater economic growth because it has invested far more in its human capital.

Physical Capital

Developing human capital alone is not enough to create economic growth. Economies must also invest in developing physical capital. **Physical capital** is the tools, factories, and equipment that are used in the production process. As the stock of physical capital increases, the nation experiences capital deepening. Capital deepening refers to the amount of capital available to each worker. Capital deepening provides for a more productive labor force. The average American worker is backed by $130,000 worth of physical capital. This is one of the reasons for America's productivity edge.

ESSENTIAL

> Some capital replaces labor, while other capital enhances it. A robot that welds automobile frames replaces workers, but a pneumatic jackhammer both enhances and replaces workers. One worker with a pneumatic jackhammer can do the work of several workers armed with sledge hammers. Regardless of whether or not it replaces or enhances, capital leads to greater productivity.

Because physical capital is the result of investment, interest rates play a key role in its development. Low, stable interest rates encourage investment. In the short run, investment creates increased aggregate demand, but in the long run it expands the economy's stock of capital. High interest rates or unstable interest rates are injurious to investment decisions and result in the formation of less capital.

Once capital is deployed it must be maintained. Capital needs adequate infrastructure to realize its potential. Roads, waterways, rail systems, and reliable utility systems make capital easier to access and greatly improve the chances that it will be used effectively. One of the failures of the Soviet Union was ineffective use of capital. The Soviets built factories that dwarfed their Western counterparts. But because of inadequate infrastructure, they were often difficult to get to. This made distributing their output more difficult. The developing world lacks steady sources of power and the transportation networks that are necessary for efficient use of factories. Europe, Japan, and the United States, by comparison, have ample infrastructure to facilitate the continuous use and transport of capital's output.

Research and Development

Creativity, innovation, and invention are necessary for economic growth to continue. Western and Japanese firms spend far more on research and development than do firms in the developing world. Research and development requires sacrificing current profits in order to gain even greater profits in the future. For firms to take this risk, incentives must exist and be protected. Patents, which provide legal protection for inventors, provide the protection firms need to realize the profits of their research and development.

If developing countries want to continue economic growth, they must find ways to encourage innovation. China's impressive growth over the last fifteen years has mainly come from the production of goods developed elsewhere. Although China has many capable, intelligent, and innovative people, the laws in China do not adequately protect intellectual property. Very little incentive exists for Chinese manufacturers to spend money on research and development of new products if the factory next door can just copy it and produce it without expending the research dollars. Over time, expect to see China make strides in protecting the intellectual property of its manufacturers.

The Rule of Law

Another condition for growth is the **rule of law**. Government officials should obey the law and should also apply the law uniformly and fairly. Corruption and cronyism discourage domestic and foreign investment by effectively raising the cost of capital. Firms, individuals, and foreign investors must know that their property is protected by law. One reason that capital investment is lacking in the developing world stems from the fact that corrupt governments are far more likely to seize private property in the name of national interests. Venezuela's seizure of foreign-owned oil fields in 2007 most likely deters future foreign investment. Unlike Venezuela, former British colonies such as the United States, Canada, Australia, New Zealand, and Hong Kong inherited the English common law with its emphasis on private property, which makes them safe and attractive to foreign investors. Foreign owners of capital have basically the same rights as domestic investors. As a result, more capital accumulates in these countries than in most others.

How Economic Policy Affects Growth

Government policies also play a role in determining economic growth. Stabilization policies by the central bank affect interest rates and thus capital investment. Fiscal policy impacts capital investment indirectly through the effect of government debt on interest rates. Tax policies that affect consumption and saving decisions influence economic growth by way of their impact on interest rates and work incentives.

FACT

A group of Chilean economists, referred to as the Chicago Boys, who studied under Milton Friedman at the University of Chicago, were instrumental in bringing market reform to the country under the dictator Augusto Pinochet. The reforms they introduced helped to halt inflation and make

Chile one of South America's fastest-growing economies. Unfortunately, the association with the dictator gave market economics a bad name on the rest of the continent.

The Fed promotes economic growth when it maintains a predictable, stable interest rate policy. Although monetary policy primarily affects short-term interest rates, it is the Fed's effect on long-term interest rates that influences growth. Firms are unlikely to make long-term investments in capital if they are uncertain about future interest rates and inflation. To avoid uncertainty, the Fed must maintain a firm lid on actual and *expected* inflation. The Fed's policy stance during a recession is to target a lower fed funds rate to encourage borrowing. However, if the Fed keeps interest rates too low for too long, future inflation is more likely. This expected inflation and increased long-term interest rates will discourage capital investment and, ultimately, long-run economic growth. Short-term increases in interest rates to dampen inflation may not please Wall Street, but since the increases reduce expected inflation, they help to keep long-term interest rates low and stable. And low, stable long-term interest rates encourage capital investment.

Fiscal policy that does not lead to a balanced budget impacts long-term interest rates and capital investment. Budget deficits in the absence of capital inflow or increased domestic saving lead to higher long-term interest rates and hinder investment in capital. If capital inflows or domestic savings are enough to offset government borrowing, the interest rate effect of a deficit is negated. Regardless of immediate interest rate effects, budget deficits, if large enough, create uncertainty and may effectively discourage investment. The presence of budget surpluses reduces long-term interest rates and encourages capital investment.

QUESTION

What is a flat tax?

A **flat tax** is one that taxes all households at the same rate regardless of the level of income. Given a flat tax of 15%, a household earning $40,000 would pay $6,000 in taxes, while a household earning $100,000 would pay $15,000 in taxes. The benefit of a flat tax is its simplicity. The downside is that for many households a flat tax would represent an increase in their tax burden. Even though many are in the 20% to 30% marginal tax brackets, their average tax rates are much lower because of exemptions, deductions, and the fact that the marginal tax rate is only on the incremental income and not the total income.

Changes in tax policy affect businesses and are likely to also impact the rate of economic growth. Increasing the tax burden on firms reduces their ability and incentive to invest in capital. Increasing the capital gains tax on financial investors reduces the flow of savings firms use to make real investments in physical capital. Businesses faced with too high a tax burden may choose to produce elsewhere. It is important to understand that capital is free to flow. Placing taxes on business, although politically popular, is a recipe for reduced growth.

Taxes on personal income affect work incentives and can thus also influence the rate of growth. In the United States, the more productive you are, the more income you earn. The more income you earn, the higher your marginal tax rate. This is what economists call a **progressive tax system**. If tax rates are increased on upper incomes, they increase the tax burden of the most productive members of society. Although American tax rates are much lower than in Europe, given a high enough tax rate, the productive worker will either reduce productivity or move to where productivity is not taxed as highly. So far, America has been the beneficiary of high tax rates in Europe. Europe has suffered a brain drain as its best and brightest, thus most highly taxed, move toward countries with lower tax rates. According to the Organisation for Economic Co-operation and Development (OECD), the brain drain is serious enough that European countries are establishing government programs to encourage expatriates to migrate back. Maybe they should try a tax cut. Europe's loss is America's gain as human capital has increased steadily in the United States.

The Downsides to Economic Growth

Economic growth is not without its down sides or its detractors. Economic growth has led to increased **income inequality**, which, if ignored, threatens continued economic growth. Over the last fifty years income inequality in the United States has increased for a variety of reasons. The loss of union power, reduction in marginal tax rates, foreign competition, and meritocracy are some frequently cited reasons. Union membership has steadily declined since the 1980s, and as a result, workers have lost leverage in negotiating wages. This decline has occurred because of structural changes in the economy and as a result of government taking a more adversarial role with unions. Decreases in marginal tax rates have also widened the gap between low- and high-income earners. According to the Census Bureau, from 1980 to 2008, the bottom quintile of households have seen little or no change in average household income, while the top has seen a steady increase in income. Globalization has contributed to income inequality. Much of this country's unskilled manufacturing has moved overseas, leaving unskilled Americans in lower-paying service sector jobs. Another theory of why income inequality has increased has to do with the development of a meritocracy. In a **meritocracy** the best and brightest marry the best and brightest and reproduce more of the best and brightest, leaving the not so best and not so bright in the proverbial dust.

For economic growth to continue, the gains must be more evenly distributed across the population. Whether that is fair or not is irrelevant. America practices both capitalism AND democracy. The more-productive members of society are rewarded by capitalism, while less-productive or less-skilled members of society see their income stagnate. This fact does not sit well with the latter and creates an opportunity for a political solution. Though not all Americans see the direct benefits of capitalism, they do have a vote. Therefore, those who wind up on the short end of capitalism's stick are likely to exercise their right to vote and change the equation. Income redistribution is a fact of life in a society with universal suffrage. Proponents of economic growth must be prepared to share their earned gains with those who may not have earned it.

CHAPTER 20: The Great Recession of 2007–20??

Before 2007, had you ever heard of subprime mortgages, CDOs, or credit default swaps? In 2007, possibly the worst financial crisis in United States' history began. Through 2008 and 2009 the financial crisis became a global recession. The scale of the financial and economic crisis is measured in tens of trillions of dollars. As this is being written, the economy has slowly begun to recover, but no one is sure if the recovery will be sustained. Hopefully, you are reading this as a history of what happened rather than as a tale of what is going on around you.

The Run-Up to the Meltdown

The **Great Recession**, as it is often called, had its beginnings in the twentieth century during a period of deregulation and rampant financial innovation. The Depression-era Glass-Steagall Act, which acted as a barrier between commercial banks and investment banks, was repealed and a new shadow banking industry was created. At the same time Fed and government regulators increasingly relied on business to regulate itself, believing that market forces would lead to self-enforcement of sound practices. While all of this was going on, the stock market was booming, leading to overconfidence by investors.

ESSENTIAL

Consumer and producer expectations are important forces in the economy. Positive expectations tend to boost economic activity while negative expectations tend to suppress economic activity. The president and the Fed Chairman are as much cheerleaders for economic optimism as they are serious policymakers.

On September 11, 2001, terrorists hijacked civilian aircraft and flew them into the World Trade Center and Pentagon while another plane crashed in a Pennsylvania field. The impact of the terrorist attacks created fear in a by-then retreating stock market. This fear helped to send the United States into a shallow recession in 2001 and 2002. The Federal Reserve responded by immediately lowering the fed funds rate and injecting large amounts of money into the banking system.

These injections, along with the Bush tax cuts, two wars, and a deregulated financial sector, led to a pool of money that created a boom in residential and commercial real estate investment. Much of the spending that occurred during the years 2002 through 2005 was fueled by home-equity borrowing at historically low rates of interest. Deregulated banks added fuel to the fire by lending money to pretty much anybody who showed up for a loan. Consumers went on a credit spending binge as the booming housing market created a wealth effect. Interest rates that should have otherwise crept up were unusually low as China, Japan, and the oil producing states continued to save large sums in the United States. Furthermore, faith in the Fed's inflation-fighting capability kept inflation fears at bay. The private, public, financial, and foreign sectors all had a hand in creating the conditions for disaster.

FACT

Asset price bubbles occur when easy credit flows toward a certain class of asset, like stocks, houses, or commodities. Precious metals are currently trading at all-time highs and may be the next bubble to burst. There is a debate among economists as to whether central bankers should raise interest rates to contain asset bubbles or allow them to run their natural course.

In 2006, the overheated housing market began to slow down. Savvy investors soon began pulling away from housing and putting money into commodities like oil and precious metals. In 2007, the housing market went into complete free-fall while oil prices shot up. This combination of events brought the spending party to a halt as consumers saw their wealth decreasing at the same time highly visible energy and food prices increased. Real estate investors, and eventually homeowners, began to walk away from properties that were now worth less than the balance on the mortgage or mortgages. Murmurs of stagflation were heard in the media. They were wrong.

Securitization

One of the culprits in the run-up to the meltdown was a financial innovation called **securitization**. Traditionally, banks made loans to customers, carried the loan on their balance sheet, and earned profit from the interest and fees. This gave banks a strong incentive to carefully assess a borrowers' risk of default. With financial innovation and shareholders hungry for ever-greater returns, more and more pressure was put on banks to increase profits by expanding their lending activities. Banks delivered these profits by becoming loan originators that charged fees to make the original loans, which they then sold to investment banks. The investment banks packaged the loans into bundles and sold them as a type of bond called a collateralized debt obligation (CDO).

CDOs were sold to institutional investors, insurance companies, other banks, and hedge funds. The investors believed that CDOs were a basically sound investment because borrowers usually pay their mortgages, and if they don't, the property is used as collateral. The popularity of CDOs expanded, which increased the size of the market. The increased popularity meant that banks could lend more and more as a ready market existed for their loans. Banks competed for customers by offering ever lower interest rates and relaxing their underwriting standards. This led to more loans being offered, and house prices continued to climb. This reinforced the market for CDOs and a vicious cycle was created that would eventually come crashing down.

ESSENTIAL

A principle-agent problem occurs when the incentives of one party come into conflict with the party they represent. In the case of CDOs, investors relied on banks to carefully underwrite the loans that were issued. However, because the banks were no longer on the hook if the loans went bad, they had little incentive to be careful in assessing a borrower's risk of default.

Risk Management and Credit Default Swaps

Many of the institutional investors, banks, and pension funds have conservative investment policies that limit the types of investments they can make. These investors rely on bond rating agencies like Moody's and Standard & Poor's to determine the overall level of risk of an investment. Many can only invest in AAA or AA+ rated bonds. These are the highest ratings and typically indicate that the investment is extremely safe. Many of the CDOs that investors bought had these high ratings.

The underlying problem was that the rating agencies were only rating the top layer of the CDO. The CDOs are divided into **tranches**, or parts, with secure, high-quality loans in the top tranche and lesser-quality loans in the lower tranches. The CDOs were packaged this way so that they would yield higher returns to the investors. The lower, riskier tranches paid higher interest rates, which made the entire CDO have a higher yield than a CDO made up of only high-quality mortgages. Once again, a principalagent problem was at work. The rating agencies earn fees from those marketing the CDOs, so they have an incentive to award high ratings to their customer's product. The institutional investors are investing other people's money, so as long as they are following the protocol and buying the highly rated investment, they have little incentive to do anything other than maximize the return to their paying customers.

To sweeten the deal, some investment banks that marketed CDOs to investors sold a type of insurance called a **credit default swap** (CDS) that would pay the investor if the CDO went into default. For the investor, this was enough to make CDOs the perfect investment. For the investment banks, they were making money hand over fist selling the CDOs and then again charging for the CDSs. Profits went through the roof, as did the incentive for managers and chief executive officers to market these products to their customers. There was only one small problem. The CDOs were much less secure than people believed and the CDSs were not adequately funded. If the CDOs were to default en masse, the investment banks that sold the CDSs would be liable for hundreds of billions of dollars. That is exactly what happened.

The Collapse of Investment Banking

Bear Stearns was the first Wall Street victim of the mortgage crisis. When the housing bubble burst, the value of the CDOs came into question. Bear had heavily marketed the CDOs and also invested in them. In the face of heavy financial losses, Bear Stearns' balance sheet became toxic. If they sold their assets, the value of the firm would plummet. Soon other investment banks refused to lend to Bear and the company faced insolvency. The New York Federal Reserve president, Tim Geithner, orchestrated a bailout of Bear Stearns by lending money to banking giant JP Morgan Chase with the understanding that JP Morgan would use the funds to purchase Bear Stearns at a deep discount. It was hoped that this would prevent a widespread panic, but exactly the opposite happened. Soon Lehman Bros. was on the ropes, but this time no one came to the rescue of the firm. The panic had spread.

As CDOs defaulted, investors exercised their CDSs, which insurance giant AIG had marketed. AIG faced losses of hundreds of billions of dollars. AIG is a key player in America's financial system, and many on Wall Street and in Washington, D.C., believed that AIG was "too big to fail." If AIG failed, the entire financial system could have possibly collapsed. The U.S. government and the Federal Reserve took the unprecedented action of bailing out AIG in order to prevent further catastrophe.

ESSENTIAL

A **moral hazard** is created when insurance or expectation of a government bailout encourages risk-taking behavior. Economists and policymakers must address moral hazard when making decisions. Failure to do so encourages more risk taking. Ignoring moral hazard is a moral hazard.

As the mortgage crisis spread, the value of CDOs, in which financial institutions had invested heavily, collapsed. With the high-profile failure of two of America's largest investment banks and the bailout of AIG, financial institutions stopped lending to each other. Banks were fearful of each other's balance sheets. When the banks stop lending to each other, banks stop lending altogether. Soon businesses not tied to housing or banking were unable to borrow, and the commercial paper market froze. Business ground to a halt. The Great Recession had begun.

A Coordinated Policy Response

The collapse of the mortgage security market meant that trillions of dollars worth of financial assets had become worthless. This economic recession was a deadly combination of a financial crisis compounded with a wider economic crisis. In the face of economic recession, the Keynesian prescription is for expansionary monetary and fiscal policy to stimulate the economy. The problem was that the tools of policy that are normally effective were both severely hampered.

The Fed

The usual prescription for a recession is for the Fed to buy treasury bills from primary security dealers and then allow the money creation process to work its magic. When banks are unwilling to lend, this prescription does not work. Instead of money being created, the financial system was destroying money. Because many of the failures in the financial system had occurred outside of traditional banking, the Fed was hard-pressed to get money where it needed to be.

The Fed is set up by law to work with the banks. The tools it has at its disposal are aimed at banks, but this financial crisis was different. The shadow banking system that had been created by deregulation was out of the scope of the Fed. In order for the Fed to get things working again, the shadow system had to be transformed into the more regulated traditional banks. That is what they did. Goldman Sachs and Morgan Stanley, the last of the great investment banks, changed their status under SEC regulations and submitted to the Fed.

Quantitative Easing

A visit to the Federal Reserve Board of Governors website throughout the financial crisis indicated that something interesting was happening. Almost on a monthly basis, the Fed was inventing new tools of monetary policy. The first was the **Term Auction Facility** (TAF), which allowed banks to bid on loans from the Fed without the perceived downside risk to borrowing from the discount window. Soon the Fed was creating other lending facilities to inject money into the system. The Fed started buying up mortgage-backed securities and agency debt. Ultimately what the Fed was doing was creating markets for assets in exchange for cash, a process called

quantitative easing. In the end, most of the facilities created during 2008 to halt the financial collapse expired in February of 2010. The Federal Reserve did not forgo traditional policy measures during the recession. The FOMC reduced its target for the fed funds rate to 0%. This is as expansionary as the Fed can get with open market operations.

Government Gets Inventive

While the Fed was busy trying to simultaneously repair a frozen banking system and return the broader economy to normalcy, the government responded by taking extraordinary measures. The treasury under Henry Paulson attempted to fix the immediate financial crisis in banking and the mortgage security market. Congress and the president, first Bush and then Obama, planned a massive policy response. The problem for government was that when the crisis hit, government was already engaged in heavy-duty fiscal expansion. Taxes were extremely low and spending was extremely high. Economists questioned how much more stimulus government could deliver.

The Congress passed the **Troubled Asset Relief Program** (TARP) in 2008, giving the treasury the ability to recapitalize the banks that had been hit hard by the collapse of the housing market and CDOs. This was only the beginning of the government's efforts to stabilize the economy. The government also purchased control of failing automaker General Motors for $50 billion to keep it from going out of business.

Another program aimed at encouraging durable goods consumption was the Car Allowance Rebate System (CARS), also known as "cash for clunkers." Under the program, buyers would receive a rebate from car dealers when trading in a used, less fuel-efficient vehicle for a new, more fuel-efficient vehicle. The dealer would then be reimbursed by the government. According to the government, 680,000 cars were sold under the program, which helped spur domestic spending in the economy. To stimulate investment in housing, the government awarded a substantial tax credit of $8,000 for homebuyers. Furthermore, unemployment benefits were extended for millions of out of work Americans. The largest of all the government's programs was the **Recovery Act of 2009**, which planned to spend almost $790 billion in order to expand the economy. Combined with tax cuts, two wars, and 0% interest rates, the stimulus is quite aggressive by historical standards. Most economists believe that the government and central bank's response to the crisis averted a second Great Depression.

Theory Versus Practice

Keynesian economic theory suggests that in the face of recession, government can run budget deficits to boost aggregate demand. The Fed can support this by engaging in expansionary monetary policy. The combined effect increases aggregate demand without creating inflation because of all the slack in the economy. This is the story that students are taught in macroeconomics, and in early 2010 it appears that is what is happening. Several questions, however, remain. Where are the jobs? Will all of this stimulus create inflation? Has government overstepped its boundaries in a capitalist society?

ESSENTIAL

Theories affect the way economists and policymakers tell the story of what is actually happening. These narratives of reality have the power to effect reality if enough people believe them. The story of expected inflation creating inflation is a powerful story, and if policymakers do not address it, the story could become the reality.

When an economy falls into recession, businesses cut production and quickly lay off workers. During economic recovery, the logical conclusion is that this process should run in reverse, but that is not necessarily the case. As production begins to pick back up, firms often discover that their remaining workforce has become much more productive. As a result, firms are able to steadily increase output without hiring more workers. As output continues to increase, firms increase the number of hours their labor force works before hiring more. Finally, after accounting for increased productivity and increased working hours, if demand is present then firms begin to hire. Unfortunately for the unemployed, this process can take a very long time. Jobless recovery is the bane of the politician's existence. The GDP may increase, but unemployment can still remain high for a while. The unemployed vote and do not really care for this neat explanation.

The Federal Reserve is walking a precarious tightrope when it comes to inflation. It does not help that so many pundits are warning of coming inflation. Expecting inflation is often enough to spark inflation. The Fed has introduced hundreds of billions in reserves to the financial system in order to unfreeze the flow of credit. The danger to the economy comes in two forms. First, if the Fed begins contracting the money supply too soon, the economy may fall back into deeper recession. Second, if the Fed waits too long to contract the money supply, then inflation may take hold, and that would require the Fed to respond by sending the economy back into recession. Ben Bernanke, the Board of Governors, and the rest of the FOMC have quite a challenge on their hands: time the withdraw of monetary stimulus so as to not spark inflation, or collapse the economy back into deep recession. Chances are they will get it right.

FACT

The Tea Party movement has emerged as a vocal critic of the fiscal and monetary policies practiced by the government and the Fed. At the heart of the movement is a distrust in the government's ability to solve economic problems. Members also decry the deficits created by expansionary policy as a threat to America's future prosperity.

The last question is probably the most important one. Did government overstep its bounds? With signs posted in yards saying "No Socialism" and the debate between Keynesians and advocates of freer markets heating up, this is an interesting time to study economics. If the economy recovers and unemployment returns to more normal levels, then the Keynesian cause will be advanced. If, however, the recession worsens in the face of stimulus or out-of-control inflation takes hold, advocates of free markets will be justified. Politicians on the right and left are becoming more vociferous in their debate over government involvement in the economy.

If the economy returns to rights soon, government involvement will have been justified. The pendulum that swings between Keynes and classical theories will stay for a while in the Keynesian world, where markets are imperfect and government intervention is a necessity. If, however, the stimulus plans and Fed interventions fail, a crushing blow will have been dealt to Keynesian economics, and you should expect to see a wave of market reforms.

CHAPTER 21: The Environment and the Economy

Most Americans believe that protecting the environment is a valuable goal for our society. The way that government and environmentalists have gone about achieving this goal has for the most part ignored the realities of economics. The Endangered Species Act, the Clean Air Act, and the Clean Water Act all have laudable goals. Critics of the legislation are not proextinction, prosmog, and pro–dirty water. For most, the criticism is in how these goals are achieved and not the goals themselves. Economists offer a unique perspective on the environment, and inclusion of economic principles can be used to bring about the goal of environmental protection more efficiently and with greater utility.

Is Growth Sustainable?

One of the costs of an ever-growing economy is the strain that it places on the environment. As the population expands more and more, resources are required to sustain the population. This growth need not necessarily lead to environmental collapse. Instead, markets can be used to alter the incentives of individuals and firms as they face trade-offs in their use of resources.

FACT

Real GDP is not the only measure of economic well-being. One that measures ecological efficiency is the **Happy Planet Index** (HPI). It measures not just material well-being but also analyzes the countries' environmental impact. America may have the largest GDP, but that does not make the planet happy according to the HPI.

Demand for resources tends to increase the price of those resources. As the price increases, individuals and firms that use these resources face an incentive to use less in the case of nonrenewables. In the case of renewable resources, entrepreneurs gain an incentive to increase the production of the renewable resource. These incentives are powerful and efficient.

Renewable Resources

Consider the case of lumber, a renewable resource. Increased demand for lumber has led to an increase in the price of lumber. This price increase leads tree farmers to expand their output to meet the demand. The net effect of increased demand for lumber is increased demand for forests. Increased demand for forests makes the land more desirable and leads to more land being placed into forest production. Some would argue that if you cut down the trees, eventually there will be none left. However, this statement ignores economic incentives. Would there be more corn or less corn if people stopped eating it? If you answered less, you would be correct. If people stop eating corn, farmers have no incentive to grow corn. Likewise, if people eat more corn, then farmers grow more corn. The same is true with trees. Trees take longer to grow, though, and as a result they're priced much higher than corn.

Nonrenewable Resources

In the case of nonrenewables like coal, oil, and natural gas, markets provide incentives to both producers and consumers. As demand increases for these factors of production, the price increases. This leads to higher prices and higher costs of production for firms that use the resources. These higher costs provide firms with an incentive to become more efficient in the use of the resource. For example, if a firm uses natural gas in production and gas prices rise, the firm has a strong incentive to use its natural gas in the most efficient way possible. Firms that are wasteful and inefficient will find it difficult to compete against firms that use resources more efficiently. They'll eventually go out of business.

Tree Huggers and Money Grubbers — Can't We All Get Along?

The goals of nature-loving people need not always be in conflict with utilitymaximizing economist types. Take the example of a fictional couple named Courtney and David. Courtney is concerned about the environment and lives her life in such a way that her ecological footprint is kept to a minimum. David, on the other hand, is not what you would call environmentally sensitive, but he is sensitive when it comes to money. Can these two get along? Yes!

ESSENTIAL

> Economic efficiency occurs as resources are used with less waste. Efficiency should be a goal of not only economists but anyone who cares about the planet. Less waste means that fewer resources are required in production.

"Greenhouse gas emissions lead to global climate change," argues Courtney. David responds, "High gas prices are killing me!" Together they decide to do something about both issues. They buy a hybrid vehicle. Courtney is happy because she now emits zero greenhouse gases. David is happy because he only spends $40.00 a month on gas.

Courtney says, "clean water is a precious resource that should not be wasted." David responds, "This water bill is killing me!" Together they decide to something about it. They install low flow toilets, dig up their grass and replace it with native drought resistant species, and decide to shower only once every other day. Both are happy, if a little smelly.

Courtney believes that "Coal mining is harmful to the environment. Burning it leads to acid deposition in our lakes and rivers, and it contributes to global climate change." David responds, "This electricity bill is killing me!" You know where this is going next. Together they better insulate their home, replace the thermostat with one that is controlled by the local utility and only comes on during off-peak hours, and replace all of their incandescent bulbs with fluorescent lights. Once again, both are happy.

Achieving positive environmental outcomes is easy when incentives become properly aligned. The same incentives that David and Courtney face could be scaled up to include industry and even nations. The environment and the economy do not have to be polar opposites. Both economic and environmental objectives can be achieved, and not necessarily at the cost of each other.

Tragedy of the Commons

Ecologist Garrett Hardin wrote about the concept of "**Tragedy of the Commons**" in a 1968 article in *Science*. According to Hardin, common areas that are not regulated by government or privately owned are subject to intense resource depletion. Hardin's basic idea is sound. If everyone owns the commons, then no one owns the commons.

For example, imagine a public pasture near a village. Farmers looking to profit by fattening their cattle allow them to graze on the public pasture. Because they do not face any private cost for this activity, their incentive is to graze as many cattle as possible on the land. Because all of the farmers face the same incentive, the land is subject to intense overgrazing and becomes useless, hence tragedy of the commons.

Several solutions exist for the problem. The most popular solution, but least effective, is government control of the commons. A more efficient solution is conversion of the commons to private property. The oldest, and sometimes most practical, solution is to let the farmers work it out among themselves.

Government control of the commons is seen in the National Park System, the Bureau of Land Management, and the National Forest Service. These agencies protect the natural resources of the nation's public lands. However, the chief decision-makers in the institutions are far removed from the commons that they administer. They often face pressure from interest groups to allow access to the commons at very little cost. This is why many of the lands under government control are experiencing the tragedy of the commons. Pastureland in the American west is subject to overgrazing as ranchers take advantage of cheap, subsidized grazing lands to fatten their herds. The National Parks, which were set aside to preserve America's natural beauty, suffer from the impact of large crowds and the necessary roads, buildings, and campgrounds to accommodate them.

FACT

Thomas Malthus was a nineteenth-century English economist who predicted that human population growth would one day exceed the food supply. Little did he know that the twentieth century would bring a transformation in agriculture, which greatly expanded the worlds' ability to produce food.

Privatizing commons is an efficient economic solution to the tragedy of the commons. If land is privately held, the owner has an incentive to maintain its productive capacity. In other words, ranchers face very little incentive to overgraze their own land. Instead, privately-held land is allowed to go fallow and herds are rotated from pasture to pasture. Ranchers on private lands face real private costs, and so they count the cost in their decision-making. This leads to more efficient outcomes both productively and allocatively.

The oldest and most practical seems to be the most commonsense solution, and that is to let the users of the commons work out a solution to administering the commons. Elinor Ostrom, winner of the 2009 Nobel Prize for economics, has studied the tragedy of the commons and concluded that local solutions among vested interests are often the best solutions. A local community is best able to regulate a commons through informal arrangements. As an example, Swiss alpine cheese makers share a grazing commons at high altitude. Their simple solution to prevent overgrazing of the commons was to only allow cattle to graze the land that had overwintered on the land. This keeps the farmers from bringing new cattle up for just the summer and overgrazing. The farmers face a cost for overwintering and so have an incentive to use the land efficiently. By the way, the Swiss came up with that solution 800 years ago. The lesson for America is this: if ranchers in Montana want to graze their cattle on a commons, then it is probably more efficient and effective for the ranchers and local community to promote a solution to overgrazing than 536 attorneys and 1,000 bureaucrats in Washington, D.C.

Human Population

One of Garrett Hardin's contentions was that the earth was a commons and that unchecked human population growth would one day lead to a depletion of the world's resources. The world population is growing at a rate of 1.14% per year. In 61 years, the world population will double from its current 6.7 billion to 13.4 billion. Hardin argued that the marginal benefit to the parent of one extra child exceeded the marginal cost because some of the cost was not borne by the parent. Many would disagree with that statement. Industrialized nations see a drop in fertility rates as women enter the workforce and become wage earners. As women become more educated they also typically delay starting a family. Where Hardin advocated government coercion to limit population, the market seems to be doing the same thing without people sacrificing their freedom to procreate.

The Role of Incentives

Protecting endangered species is important to many people. There is considerable political pressure for governments to enact legislation to protect species. One way to protect endangered species that is promoted by economists is to kill them for food.

Why did the American bison population almost go extinct while the cow population increased exponentially? Is it because the cows ate the bison? No, the reason bison faced extinction while cows thrived is because cows were private property while bison were a commons. The population of bison fell from the millions to just over 1,000 by 1889. Today, fortunately, the American bison is back from the brink of extinction because they ceased to be a commons and became private property. Today there are more than 500,000 bison, and their numbers are growing as a market has developed for bison meat.

In Africa, some countries have banned the hunting of elephants, while in others elephants are hunted for food and sport. In which countries is the elephant population declining and in which is the elephant population actually growing? In Kenya, Tanzania, and Uganda, where elephant hunting is banned, there is little incentive to raise and breed elephants, but there is an incentive to poach them for their ivory. Elephant poaching is a dangerous business, and those who do it are heavily armed. Park rangers paid to protect the elephants are often outgunned and outmanned, which leads to not only loss of the elephants but also the loss of human life. But in Malawi, Namibia, and Botswana, elephant populations are growing and thriving as commercial hunting revenue provides a strong incentive for breeding the elephants. The revenue from commercial hunting activity is also used to protect habitat for the species. Many would-be poachers are employed in maintaining and protecting these countries' growing elephant populations.

The Right to Pollute

Pollution is an economic bad. All production processes create some form of pollution, so zero pollution is not a reasonable goal. What is the right amount of pollution? The right amount, or **socially optimal** amount, of pollution occurs when the marginal social benefit equals the marginal social cost of production. That means that firms should produce to the point where the extra benefit of production to society equals the extra cost, inclusive of the cost of pollution to society. Because of tragedy of the commons, firms often do not have an incentive to produce the amount of output that is socially optimal. Firms do not pay the cost of pollution, and so they produce too much output and thus too much pollution. In order to get firms to produce the appropriate amount of pollution, the cost of pollution must enter into their production decision. Governments can tax or sell pollution permits. Another option is for affected individuals who bear the cost of the pollution to negotiate a payment from the polluter.

ESSENTIAL

There are really no such things as clean air or clean water. There is only air and water with varying amounts of pollutants in them. The question is, how much are you willing to put up with? To deal with environmental problems, people must be realistic and reasonable.

Per Unit Taxes

A per unit tax on the production of a good or service could be used to reduce the amount of pollution that the firm produces. The tax on the producer increases the cost of production, which reduces their willingness and ability to produce their product. The market outcome is for the price of the good or service to increase and for the quantity to decrease. The result is less production and, therefore, less pollution. The problem with a per unit tax is that it would most likely be levied on all producers in an industry, which means that cleaner, more efficient producers are taxed at the same rate as the heavier polluters. The tax reduces pollution in the industry but does not increase the incentive for individual producers to clean up their act. Also, if demand for the good or service increased, then the quantity of pollution would still increase. The ultimate goal is to reduce pollution and not just punish producers.

Pollution Permits

Another option is to create a market for pollution permits. Under a permit scheme, the government establishes a cap on how much of a certain pollutant will be allowed in the atmosphere that year. For example, assume that last year seven million tons of nitrogen oxide gas was released into the atmosphere and government wants to reduce it to six million tons. Government would allocate to industry the number of permits required to achieve this goal. If each permit allowed one ton of emissions, then government would allocate six million permits to firms. The firms would be required to surrender one permit for each ton of pollutant they produced. Under the system, firms and even individuals could buy and sell the permits in an exchange. Firms that are relatively clean could sell their unused permits to firms that are heavier polluters. Individuals could also buy permits and effectively keep a certain amount of pollution out of the atmosphere. This scheme encourages firms to become more efficient and reduce pollution simultaneously. Unlike pollution taxes, this solution rewards firms for reducing pollution instead of punishing all firms equally.

FACT

Climate exchanges exist worldwide for trading permits in various industrial pollutants. Climate exchanges act much like stock exchanges allowing firms to buy and sell permits. In the future, students interested in the environment and economics may become climate brokers.

The Coase Theorem

A final option for firms that pollute or harm the environment is to directly pay those who are harmed by the pollution. This option is referred to as the **Coase theorem**. Ronald Coase, a British economist, suggested that an efficient outcome could be achieved if polluters and those who bear the cost directly negotiated a payment that was acceptable to both parties. Some assumptions of the Coase theorem are that there are no bargaining costs for either party, property rights are clearly defined, and the number of people involved is small. An application of the Coase theorem could be applied to the Dr. Seuss classic book *The Lorax*. In the book, a greedy capitalist named the Once-ler busily chops down trufula trees in order to make thneeds. The Lorax, the official spokesman of trees, rages against the deforestation at the hands of the Once-ler. Because of tragedy of the commons, the Once-ler never counts the cost of the trufula trees and by the end of the story has wiped them out. If clear property rights had been assigned, then this tragedy would not have occurred. If government had ceded private property rights to either the Once-ler or the Lorax, then the two could have negotiated a payment for trufulas that would have satisfied both parties.

The Nature Conservancy and Sierra Club use this strategy to preserve environmentally sensitive areas. Many concerned with the environment have applied this strategy in preserving the rain forest. It is much easier to save a forest that you own than to save a forest that everyone owns. Economic principles can be used to effectively and efficiently achieve environmental goals.

APPENDIX A

Economics Glossary

absolute advantage

This exists when an individual or a nation can produce goods and services either quicker or in greater abundance than can others. Absolute advantage implies greater efficiency.

acceptability

The condition that exists when people readily accept some form of money as a means of payment.

aggregate demand

The inverse relationship that exists between the real GDP and the price level in the economy. Aggregate demand (AD) is the willingness and ability of households, firms, government, and the foreign sector to buy the new, domestic production of a nation.

aggregate supply

The real gross domestic product that firms are willing to produce at each and every price level. In the short run, real GDP and price level are directly related, but in the long run, real GDP is independent of the price level.

allocation

The act of getting resources to where they are needed or wanted most.

allocative efficiency

The condition where marginal benefit equals marginal cost.

Animal Spirits

Keynes's term to describe his contention that people's actions are motivated at times by fear or hubris rather than reason.

arbitrage

Taking advantage of different prices in different markets in order to profit.

assets

Everything that is owned by an individual or firm. For banks, assets include its reserves, loans to customers, securities, and real estate.

autonomous consumption

The level of household spending that occurs independent of disposable income. Autonomous consumption does not vary with changes in the business cycle.

balance of trade deficit/surplus

The degree to which either exports or imports exceed each other. When exports exceed imports, a balance of trade surplus exists. If imports exceed exports, then a balance of trade deficit exists.

bank balance sheet

Comparison of a bank's assets to its liabilities and equity. Changes in the balance sheet influence the bank's ability to lend.

bank run

A situation where fact or rumor motivates bank customers to immediately demand their funds from their bank. Because banks only hold a fraction of their reserves as cash, the first to demand their funds are usually the only ones who walk away with money.

barter

Exchanging one good or service for another good or service.

Beige Book

A survey of anecdotal economic information gathered by each of the Federal Reserve's twelve district banks that is used by the FOMC in making interest rate decisions.

black market

Term used to describe a market where illegal goods or services are exchanged.

Board of Governors

The government part of the Federal Reserve System. Members of the Board of Governors are appointed by the president of the United States and are confirmed by the Senate. The Board of Governors is responsible for making the rules and regulations for the Federal Reserve System. The chairman of the board is appointed every four years and serves as the Federal Reserve System's leader as well as representing the United States's economic interests with the International Monetary Fund and the World Bank.

bonds

A security that is a promise from a borrower to pay a lender on a specified date with interest.

budget constraint

The amount of disposable income available to consumers for purchasing goods and services.

business cycle

Changes in the real GDP over time. The real GDP has periods of sustained increase punctuated with brief contractions also known as recessions.

capital

Capital in economics does not refer to money but to all of the tools, factories, and equipment used in the production process. Capital is the product of investment.

capital controls

Restrictions on the inflow and outflow of foreign investment in a country.

capital deepening

The process by which the amount of tools, factories, and equipment increases relative to the size of the labor force so that each worker has more available capital with which to work.

capital gain

An increase in the price of a real or financial asset. For example, if you buy Coca-Cola stock at $30.00 per share and then sell it at $50.00 a share, then you earned a $20.00 per share capital gain.

capital requirements

These ensure that banks are able pay depositors if some of the bank's borrowers are unable to repay their loans. Banks are required by law to maintain capital requirements.

capitalism

A hybrid economic system that emphasizes markets as the best means of allocating goods and services while recognizing a role for government as a provider of public goods and as a regulator.

cartel

A group of producers that agree to cooperate instead of compete with each other. Cartels seek higher profits for their members by collectively reducing production in order to increase prices.

ceteris paribus

When studying the relationship between different economic variables, economists hold all variables constant except for what is being studied. The assumption allows economists to find the relationships between different variables.

classical economics

The school of economic thought that assumes that market forces are able to efficiently and effectively allow an economy to remain fully employed. Freedom from government intervention in the economy is a logical conclusion of classical economics.

Coase theorem

Ronald Coase suggested that an efficient outcome could be achieved if polluters and those who bear the cost directly negotiated a payment that was acceptable to both parties. Some assumptions of the Coase theorem are that there are no bargaining costs for either party, property rights are clearly defined, and the number of people involved is small.

COLA

COLA stands for cost of living adjustment. Social Security recipients in addition to many pensioners receive COLA to offset the effects of inflation on their income.

collateralized debt obligation (CDO)

A packaged bundle of mortgage loans sold as a security similar to a bond. By bundling different loans together, investment banks sought to diversify the risk that investors faced.

collusion

This occurs when businesses illegally cooperate in order to make higher profits.

command economic system

The key characteristic of the command economy is centralized decision-making. Either one or a group of powerful individuals makes the key economic decisions for the entire society.

commodity money

When relatively scarce minerals, metals, or agricultural products are used as a means of exchange.

common stock

A security that provides shareholders with ownership rights in a corporation. Common stockholders are able to vote for the board of directors.

comparative advantage

Exists if you can produce a good at a lower opportunity cost than someone else. In other words, if you sacrifice less of one good or service to produce another good or service, then you have a comparative advantage.

complements

Goods that are consumed in conjunction with each other like chips and salsa. When the price of a complement increases or decreases the demand for the other complement moves in the other direction. For example, if the price of salsa rises, then the demand for chips falls. Economists calculate cross-price elasticity to determine whether or not goods are complements or substitutes.

consumer price index (CPI)

The consumer price index is a market basket approach to measuring inflation constructed by the Bureau of Labor Statistics. Changes in the CPI indicate inflation or deflation. CPI is the most widely used measure of inflation in the American economy.

consumption

Household spending on new domestic goods and services. Consumption accounts for more than two-thirds of the gross domestic product.

contractionary fiscal policy

Government efforts to stabilize prices by raising taxes and/or reducing government spending.

contractionary monetary policy

Central bank efforts to reduce the size of the money supply of a nation.

contractionary policy

Policy conducted by either government or the central bank that seeks to limit and control inflation. The government achieves contractionary policy by reducing spending and/or raising taxes. The central bank enacts contractionary policy by reducing the money supply in order to raise interest rates, which discourages borrowing and encourages saving.

copyright

Legal protection for intellectual property. A copyright allows the author of an original work to exercise a legal monopoly over their intellectual property.

corporate bonds

Bonds issued by businesses in order to raise the financial capital with which to invest in physical capital.

cost-push inflation

Inflation caused by a decrease in aggregate supply. Unlike demandpull inflation, cost-push inflation is often accompanied with higher unemployment. The condition known as stagflation is associated with cost-push inflation. Cost-push inflation tends to be self-limiting as the concurrent high unemployment eventually brings prices down.

coupon bond

Bonds that are sold at or near face value and provide guaranteed interest payments.

creative destruction

A term coined by economist Joseph Schumpeter that refers to the ongoing process of technological innovation and industrial decline. As one industry is being born, another industry is dying. The death of old industry frees up the land, labor, capital, and entrepreneurship that can now be employed in the new industry. For example, as the car industry took off, the horse and buggy industry fell into decline. The resources once employed in the old industry could now be employed in the automobile industry.

credit default swap

Also referred to as a CDS, credit default swaps are a type of insurance on collateralized debt obligations that promise to pay the investor if the CDO should go into default. The government's takeover of AIG is tied to the risk the company took by issuing billions in credit default swaps that were exercised during the 2007–2008 financial crisis.

cross-price elasticity

The rate of change in quantity demanded of a good divided by the rate of change in the price of a related good. Positive cross-price elasticity indicates that goods are substitutes, while negative cross-price elasticity indicates that goods are complementary.

cyclical unemployment

Unemployment associated with downturns in the business cycle. Most economists view cyclical unemployment as harmful and believe that government intervention is necessary to prevent it from occurring.

deadweight loss

The loss of consumer surplus and producer surplus that results when a price other than the market equilibrium price occurs.

default risk premium

A premium added to the interest rate on a bond that has a risk of default. The greater the chance of default, the higher the default risk premium.

deflation

A general decrease in prices as a result of a decline in gross domestic product. Deflation is harmful to the economy because it creates an incentive for consumers and investors to delay their purchases. This results in even lower prices, thus creating a negative feedback loop.

demand

The willingness and ability to buy a good or service in a given time period.

demand-pull inflation

Inflation that results from increases in aggregate demand. Demand-pull inflation persists if the monetary authority increases the money supply.

depreciation

As capital ages, its value declines because it breaks down and eventually needs replacement. Depreciation expense is incurred by businesses in order to replace worn capital with new capital. Depreciation = Gross Investment − Net Investment.

diminishing marginal utility

Each additional unit of a good or service consumed gives less utility than the previous unit. Diminishing marginal utility is central to the law of demand, which says that the price of a good varies inversely with the quantity demanded. For example, given an entire pizza, each additional slice consumed gives less satisfaction than the previous slice. As a result, the price of a whole pizza is less than the sum of the individual slice price.

diminishing returns

The law of diminishing returns states that initially adding inputs to the production process yields increasing amounts of output. However, after a certain point, adding inputs increases output at a diminishing rate until eventually adding inputs actually decreases output.

discount rate

The discount rate is the nominal interest rate that the Federal Reserve charges member banks on overnight loans. Because most banks borrow from each other in the fed funds market, the primary purpose of the discount rate is to signal future changes in the more important fed funds rate.

discouraged workers

People who, for whatever reason, have given up the job search and are not officially classified as unemployed. The presence of discouraged workers in the economy means that the official unemployment rate understates actual unemployment.

disinflation

A decline in the rate of inflation. Disinflation or moderation is a beneficial side effect of increases in productivity and well-managed inflation expectations.

disposable income

Income available to consumers after they have paid taxes. Disposable income is a key determinant of consumption and saving.

dividends

Divided shares of a corporation's profits. If a corporation makes a profit, the board of directors may decide to declare a dividend on a per share basis.

divisibility

The characteristic of money, which means that it can be broken down into smaller units. For example, dollars can be broken down into quarters, dimes, nickels, and pennies.

dominant strategy

A strategy that is the player's best, independent of another player's strategy.

durability

The characteristic of money that is met when money stands up to frequent use.

economic growth

A sustained increase in real GDP over time. Economic growth occurs when the quality or quantity of an economy's productive resources increases.

economics

The study of how individuals, institutions, and society choose to deal with the condition of scarcity.

economies of scale

As the size of a firm's physical plant increases, the average cost of production initially decreases. Economies of scale explain why larger firms can often produce goods and services at lower average cost than can smaller firms.

efficiency wage

A wage that exceeds the market wage. Efficiency wages encourage worker productivity but also play a role in creating unemployment.

elasticity of demand

The sensitivity of the quantity demanded for a good to changes in the price of the good. It is calculated as the rate of change in quantity demanded divided by the rate of change in the price. Elasticity greater than one indicates demand is elastic. Elasticity less than one indicates demand is inelastic.

elasticity of supply

The sensitivity of the quantity supplied for a good to changes in the price of the good. It is calculated as the rate of change in quantity supplied divided by the rate of change in the price. Elasticity greater than one indicates supply is elastic. Supply elasticity of less than one indicates inelastic supply.

embargo

An embargo is a ban on trade with another country. The purpose of an embargo is usually to punish a country for some offense.

employed

The condition of having a job.

employment-to-population ratio

The number of employed people divided by the working age population.

entrepreneurship

The abillity to create new businesses by combining land, labor, and capital in new ways to provide a good or service. Entrepreneurs are unique individuals who are willing to take great risks. These people are willing to risk their wealth in order to earn profits.

equilibrium

The condition that exists when the quantity supplied equals the quantity demanded at one price in the market.

excess reserves

Bank reserves that are available for lending. When customers make checkable deposits, a small percentage must be set aside as required reserve and is unavailable for lending. The remainder is excess reserves and is fully loanable.

exchange rate

The price of a single unit of currency in terms of another currency. The exchange rate is subject to the supply and demand for a nation's currency.

excludable good

A good is considered excludable if it can be withheld if it is not purchased. An example of an excludable good is a fast-food meal. If the customer does not pay for it, then they don't receive it.

expansionary fiscal policy

Government efforts to return the economy to full employment by lowering taxes and/or increasing government spending.

expansionary monetary policy

Central bank efforts to offset the upside pressure on interest rates by expanding the money supply and lowering short-term interest rates. This allows for fiscal policy to be more effective.

expansionary policy

Fiscal or monetary policy targeted to increase aggregate demand in order to increase real GDP and reduce the unemployment rate. Government can increase spending or reduce taxes to achieve this goal while the central bank can increase the money supply, thus reducing nominal interest rates to encourage investment.

expected inflation

The generally anticipated inflation rate in the economy. Savers and lenders offset inflation's effects by adding the expected inflation rate to the real interest rate that they would like to earn.

fed funds

Private bank reserves that are on deposit with a Federal Reserve district bank. These reserves are loaned and borrowed in the fed funds market.

fed funds market

The market where fed funds are exchanged so that bank reserve requirements can be met and where banks with excess reserves can lend to banks in need of more reserves.

fed funds rate

The nominal interest rate that banks charge each other for the overnight use of fed funds. The federal funds rate is targeted by the Federal Reserve and is a key interest rate in the economy.

Federal Reserve System

The central bank for the United States. The Federal Reserve System is made up of both the private and public sector. The private sector is represented by the Federal Reserve's twelve district banks while the public sector is represented by the Board of Governors in Washington, D.C. Unlike most central banks, the Federal Reserve System is decentralized, and no single district bank is superior to another.

financial markets

In the circular flow model, financial markets are where savings are exchanged for financial assets.

financial sector

The financial sector of the economy is represented by banks, insurance companies, brokerages, exchanges, credit unions, and other banklike institutions that link savers with borrowers. The financial sector is an intermediary in most transactions between the private, public, and foreign sectors.

fiscal policy

The use of the federal budget in order to reduce unemployment or stabilize prices.

fixed cost

Costs that do not change with a firm's level of output. Examples of fixed costs include rent, property taxes, management salaries, and loan payments. Fixed costs are often referred to as overhead.

fixed exchange rate

An exchange rate that is managed by a government or central bank so that it does not vary. Maintaining a fixed exchange rate requires that the government or central bank actively buy and sell the nation's currency in order to stabilize the rate.

flat tax

A tax that taxes all households at the same rate regardless of the level of income.

floating exchange rate

A floating exchange rate is one that is subject to the forces of supply and demand. The United States dollar floats against other currencies.

FOMC

The Federal Open Market Committee is the chief decision-making body of the Federal Reserve System. It is composed of the Board of Governors and the presidents of the district banks. The FOMC meets eight times a year in order to evaluate the economy and make any necessary changes to the nation's monetary policy. The FOMC's voting members include the Board of Governors, the New York Federal Reserve Bank president, and four of the remaining eleven presidents who rotate on and off the committee every year. The FOMC's decisions are reflected in changes in the fed funds rate.

foreign factor income/payment

Rent, wages, interest, and profits that flow in and out of a country as the land, labor, capital, and entrepreneurship are exported and imported from and to a country.

foreign sector

The rest of the world relative to a country is considered the foreign sector.

free trade

Trade that is based on mutually beneficial exchange and that is not distorted by government intervention.

free-rider problem

The problem that occurs when a good is nonrival and nonexcludable. Some people are able to acquire the good or service without paying for it. For example, listeners of public radio that do not pay for it but still listen to it are considered freeriders.

frictional unemployment

Voluntary unemployment that occurs when a person enters the labor force and looks for a job. A recent graduate looking for her first job out of college would be considered frictionally unemployed. Frictional unemployment always exists in an economy.

full employment

The level of employment that exists when the economy is being productively efficient. Full employment is associated with an economy at the natural rate of unemployment.

game theory

A study of the outcomes of the decisions made when those decisions depend upon the choices of others. In other words, game theory is a study of interdependent decision-making.

General Agreement on Tariffs and Trade

In 1947, many nations, including the United States, came together and formed the General Agreement on Tariffs and Trade, or GATT. The goal of GATT was to reduce tariffs and other trade barriers so that member countries could equally enjoy the benefits of free trade. Through a series of negotiations, or rounds, often lasting years, tariffs were significantly reduced and international trade expanded in the latter half of the twentieth century.

gold standard

Under a gold standard, money represents and is convertible to a fixed amount of gold. This limits the ability of the government to print currency.

government monopoly

A government monopoly is said to exist when the government is the sole producer of a good or service. In many nations, the country's airline is owned by the government. This represents a government monopoly.

government spending

Government purchases of final domestically produced goods and services and also wages paid to government employees are included in the definition of government spending. Transfer payments like Social Security are not included in this component of GDP.

gross domestic product (GDP)

The total value of all final domestic production in a country for a given year. In addition, GDP is a measure of all the spending and all of the income spent and earned in producing the domestic output.

gross private investment

Firms and households spending on new physical capital, new home construction, and new business inventories produced domestically.

Happy Planet Index

An alternative measure of an economy's health that accounts for not only the production of goods and services but also environmental impact.

Heckscher-Ohlin theory

A theory that if two countries have different mixes of labor and capital and specialize according to their mix, then they will benefit more from trade than two countries with similar mixes of labor and capital.

Herfindahl-Hirshcman index

A tool for measuring the degree of market concentration in an industry. Low index numbers indicate that an industry is relatively competitive, while high index numbers are indicative of oligopoly.

horizontal merger

This occurs when businesses in the same stage of production combine to form a single firm. If Kroger were to combine with Winn Dixie, then a horizontal merger would have occurred in the grocery business.

human capital

The education, skills, and abilities possessed by an individual.

hyperinflation

Hyperinflation is informally defined as a very high rate of inflation. Some definitions state that hyperinflation is monthly inflation greater than 50%. Historical rates of hyperinflation have been exponential. Generally speaking, hyperinflation is inflation that is bad enough for people to refer to it as something other than just inflation.

imperfect competition

The condition that exists in a market when not all of the conditions for perfect competition are met. For example, perfect competition assumes that producers deal in identical products, but monopolistically competitive firms engage in product differentiation and thus exist in imperfectly competitive markets.

incidence of taxation

This determines who bears the burden of a tax. Statutory incidence is the legal incidence of who pays the tax. Economic incidence describes who actually pays the tax when the elasticity of supply and demand is taken into account.

income effect

When the price of a good rises, the purchasing power of a consumer's income falls, and as the price of a good falls, the consumer's purchasing power rises. The income effect is used to explain why the law of demand exists.

income elasticity of demand

The rate of change in the quantity demanded of a good divided by the rate of change in consumer's income. Income elasticities greater than one indicate normal luxuries, income elasticities between zero and one indicate normal necessities, and income elasticities less than zero indicate that a good is inferior.

income inequality

The uneven distribution of income among households in a country.

inconvertible fiat

Refers to both paper and virtual money that is intrinsically worthless and is not redeemable or backed by some real commodity.

increasing returns

The stage of production in which adding inputs increases both total product and marginal product.

inferior good

An inferior good is one where the income elasticity of demand is less than zero. In other words, as income increases, demand for inferior goods decrease. An example of an inferior good might be generic or store brand macaroni and cheese.

inflation

A general increase in the overall price level or a decrease in the purchasing power of money.

initial public offering

The initial sale of shares in a company in order for the company to acquire the necessary financial capital it needs to expand by investing in physical capital.

interest

A payment for using money.

interest rate

The price of using money expressed as a rate.

interest rate parity

The difference between interest rates in countries, eventually reflected by the difference in the exchange rate.

International Monetary Fund

An international financial organization created to provide financing and expertise for both developed and developing countries.

international trade

The exchange of goods and services between individuals and firms in different countries.

investment

See gross private investment.

Keynesian economics

A school of economic thought that emphasizes the need for government intervention in order for the economy to function efficiently and remain fully employed. Keynesian economics is based on the work of the twentieth-century British economist John Maynard Keynes.

labor

Labor refers to people with their skills and abilities. Labor is divided into unskilled, skilled, and professional. Unskilled labor refers to people without formal training that are paid wages to do repetitive tasks such as making hamburgers or performing assembly line production. Skilled labor refers to people paid wages for what they know and what they can do. Welders, electricians, plumbers, mechanics, and carpenters are examples of skilled labor. Professional labor is paid wages for what they know. Doctors, lawyers, engineers, scientists, and even teachers are included in this category.

labor force

The number of employed plus unemployed people age sixteen and over.

labor force participation rate

The labor force divided by the working age population.

Laffer curve

The relationship between tax rates and tax revenue expressed by economist Arthur Laffer. The Laffer curve shows that, at first, increasing tax rates raises tax revenue but eventually reduces tax revenues as tax rates continue to increase.

laissez faire

The philosophy that the economy functions better when government takes a hands-off approach to economic matters.

land

Land is inclusive of all natural resources and not just some random piece of property. Trees, mineral deposits, fish in the ocean, ground water, and plain old land are all included. Land can be divided into renewable and nonrenewable natural resources.

law of demand

The inverse relationship that exists between price and quantity. The law of demand is attributable to income effect, substitution effect, and the law of diminishing marginal utility. The fact that more people are willing to buy at lower prices than at higher prices is called the law of demand.

law of one price

Also known as purchasing power parity; after accounting for the exchange rate, the prices of similar goods should be the same regardless of where they are purchased.

law of supply

Price is directly related to quantity. As prices increase, quantity supplied increases, and vice versa. The law of supply states that producers are able and willing to sell more as the price increases. As production increases, so do the marginal costs. As rational, self-interested individuals, suppliers are only willing to produce if they are able to cover their cost.

legal tender

Currency that must be accepted as payment for a debt.

liabilities

Debts and other payable obligations. With regards to bank balance sheets, the liabilities include customer deposits and loans from other banks and the Fed.

liquidity preference theory

A theory of interest rate determination, which says that the nominal interest rate is a function of the money supply, and explains people's preference for holding cash instead of other financial assets.

liquidity premium

A premium added to the interest rate on a bond that might be difficult to sell in a secondary market.

liquidity trap

Because nominal interest rates cannot be reduced below zero, central banks are unable to induce more investment and spending once 0% nominal interest is reached.

loanable funds market

The market that brings together savers and borrowers. The equilibrium of saving and borrowing determines the real rate of interest.

long run

In microeconomics, the long run is the period of time in which all costs are variable. In macroeconomics, the long run is the period in which input prices adjust to changes in the price level.

long-run aggregate supply (LRAS)

The long-run aggregate supply shows that the level of real GDP is independent of the price level when input prices are able to adjust to changes in the price level.

long-run equilibrium

A long-run equilibrium occurs when aggregate demand intersects the short-run aggregate supply at full employment. A long-run equilibrium is characterized by stable prices and inflation expectations.

long-run Phillips

curve The long-run Phillips curve shows that in the long run, the natural rate of unemployment is independent of the level of inflation. Compared to short-run Phillips curves, no trade-off exists between the rate of inflation and the natural rate of unemployment.

M1

All of the checking account balances, cash, coins, and traveler's checks circulating in the economy.

M2

Everything in the M1 plus savings account balances, certificates of deposit, money market account balances, and U.S. dollars on deposit in foreign banks.

macroeconomic equilibrium

The intersection of aggregate demand and short-run aggregate supply, which determines the price level and the current level of real GDP.

macroeconomics

The study of how entire nations deal with scarcity. Macroeconomists analyze the systems nations create or allow to allocate goods and services. The questions they ask are varied and of great interest to individuals and policymakers alike. How do you measure the economy? Why does unemployment exist? How do changes in the amount of money affect the entire economy? What impact does government spending or tax policy have on the economy? How can you make the economy grow?

marginal benefit

Because people usually make decisions one at a time, economists refer to the benefit of a decision as marginal benefit.

marginal cost

The cost incurred to either produce or consume one extra unit of whatever it is you are producing or consuming.

marginal product

The additional output produced by the addition of extra inputs. For example, if adding a worker to an assembly line increases car production by three units, then the marginal product of the additional worker is three cars.

marginal revenue

The additional revenue earned by a firm for producing one more of a good or service.

marginally attached workers

These are people ready and available to work who have conducted a job search within the past twelve months but have not searched in the last four weeks and are therefore not included in official unemployment statistics.

market

Any place that brings together buyers and sellers. Most economists believe that markets are the most efficient means of allocating scarce goods and services.

market clearing price

The market equilibrium price where the quantity supplied equals the quantity demanded. At the market clearing price, neither surpluses nor shortages develop.

market economy

Market economies are characterized by a complete lack of centralized decision-making. As opposed to top-down planning, market economies operate bottom-up. Individuals trying to satisfy their own self-interest answer the questions of what, how, and for whom to produce. Instead of tradition or government mandate, the "invisible hand" of the market directs resources to their best use. Private citizens, acting on their own free will as buyers or sellers, trade their resources or finished products in the market in order to increase their own well-being.

market failure

This exists when the market fails to provide the allocatively efficient amount of a good or service. Also, market failures exist whenever a third party is either positively or negatively influenced by production.

maturity risk

A premium added to a bond to offset the probability that it will be refinanced if interest rates drop.

medium of exchange

This means that money is being used for the purpose of buying and selling goods or services.

mercantilism

The idea that a country and, therefore, individuals, are better off if the value of a country's exports are greater than the value of its imports. Under mercantilism, the more gold a country amassed, the wealthier it became. As a result, countries competed to import cheap natural resources and then convert them into more expensive manufactured goods for export.

meritocracy

The theory that success and achievement based on merit has led to unequal income distribution. The theory suggests that those with more human capital tend to associate and marry others with similar human capital, resulting in a fragmented society of haves and have-nots based on merit rather than chance.

microeconomics

Economics that focuses its attention on the decision-making of individuals and businesses. Microeconomics is primarily concerned with markets for goods, services, and resources.

misery index

An index of economic not-so-well-being that is found by adding the inflation rate to the unemployment rate. A normal result is 7%. The idea was postulated by economist Arthur Okun.

monetary policy

Efforts by the central bank of a country to stabilize prices, promote full employment, and encourage long-run economic growth through controlling the money supply and interest rates. In the United States monetary policy is conducted by the Fed.

monetizing the debt

This occurs when the central bank purchases the newly issued debt of the government, thus exchanging cash for bonds. Monetizing the debt is analagous to printing up money to pay debts and is highly inflationary.

money

Money is anything that functions as a medium of exchange, store of value, or standard of value.

money market

The market where the central bank supplies money and the other sectors of the economy demand money. The money market determines the nominal interest rate.

money multiplier

The degree to which changes in bank excess reserves result in changes in the money supply. The money multiplier can be calculated by dividing one by the required reserve ratio.

monopolistic competition

A form of imperfect competition in which otherwise competitive firms engage in product differentiation in order to earn economic profits.

monopoly

A form of imperfect competition characterized by one producer of a good or service.

monopsony

The condition of having one buyer in a market. A single buyer has significant influence over sellers in the market when it comes to price negotiation.

moral hazard

The idea that the presence of insurance or government assistance encourages unnecessary risk-taking behavior on the part of economic actors.

multiple deposit expansion

The process by which money is created when banks accept customer deposits and then lend excess reserves, resulting in an expansion of demand deposits in the banking system.

municipal bonds

Bonds issued by local governments used to pay for capital and infrastructure improvements. Municipal bonds are exempt from federal income taxes and are thus popular among bond investors.

natural monopoly

A monopoly that exists because of economies of scale. Natural monopolies are more efficient than several smaller competitors.

natural rate hypothesis

See long-run Phillips curve.

natural rate of unemployment

The rate of unemployment that exists when there is no cyclical unemployment present in the economy. The natural rate of unemployment is thought to be independent of the inflation rate.

negative externality

A negative externality exists if a third party bears part of the cost of production. For example, a paper mill that pollutes a local river creates a cost for local residents that is not reflected in the price of the paper being produced.

negative returns

In the production function, this is the stage of production in which total product and marginal product are both decreasing.

net exports

Exports minus imports. Net exports is also referred to as the balance of trade. When net exports are positive, a trade surplus exists, but when net exports are negative, then a trade deficit is said to exist.

net foreign investment

The balance of direct investment between countries. Net foreign investment includes both real and financial investment.

net investment

The purchase of new capital that expands the economy's capacity to produce. Net investment = Gross private investment – Depreciation.

nominal GDP

The gross domestic product measured in current dollars. Nominal GDP can be found by multiplying the real GDP by the price level.

nominal income

Income measured in current dollars.

nominal interest rate

The nominal interest rate is equal to the real interest rate plus expected inflation plus other premiums for risk. Nominal interest rates are the ones that you see at a bank or in the news.

nonexcludable

A good or service is nonexcludable if it cannot be withheld if a user does not pay for it. Public schools are an example of a nonexcludable service.

nonprice competition

Competition based on something other than price. Nonprice competition may involve product differentiation or advertising.

nonrival

Goods and services are considered nonrival if one person's consumption of the good or service does not diminish another's ability to consume the same good or service. Movies are nonrival. The presence of other moviegoers does not diminish your consumption of the good.

normal good

A normal good is one with income elasticity greater than zero. As income increases, demand for normal goods increases, and vice versa.

normal luxuries

A normal good with income elasticities greater than one. A good is a normal luxury when the rate of increase in the quantity demanded exceeds the rate of income growth. Normal luxuries might include foreign travel, expensive restaurant meals, or German sports cars.

normal necessity

A normal necessity is a good with income elasticity between zero and one. As income increases, the demand for normal necessities increases at a lower rate. For example, as income increases, the demand for toothpaste increases at a lower rate than the change in income.

North American Free Trade Agreement (NAFTA)

A treaty that reduced trade barriers between Canada, the United States, and Mexico.

official reserves

The foreign currency holdings of a central bank. Official reserves are used by some central banks to stabilize the exchange rate.

Okun's law

For every 1% that the actual unemployment rate exceeds the natural unemployment rate, a 2.5% decline in real GDP is anticipated.

oligopoly

A market condition that exists when a few large sellers dominate the market for a good or service. Most large industries in America are oligopolistic.

open market operations (OMO)

Open market operations occur when the New York Federal Reserve engages in buying or selling treasury securities with primary security dealers in order to influence the money supply and the fed funds rate.

opportunity cost

The next best alternative use of a resource.

Organization of Petroleum Exporting Countries (OPEC)

A cartel of oil-producing nations that seeks to increase profits for its members by collectively limiting production.

overhead

See fixed cost.

patent

Legal protection for an invention or process that gives the holder a monopoly for whatever it is that was patented.

perfect competition

An industry characterized by a large number of buyers and sellers, each acting independently according to their own self-interest, perfect information about what is being traded, and freedom of entry and exit to and from the market. Firms deal in identical products and they are "price takers."

perfect information

The condition in perfect competition where buyers and sellers have complete information about what is being bought and sold, and make their decisions based on all of the information.

personal consumption expenditure (PCE)

See consumption.

personal consumption expenditure deflator

A price index used to measure consumer price inflation that is more inclusive than the popular CPI.

Phillips curve

The trade-off between inflation and unemployment that was observed both in Britain and the United States. *See also* short-run Phillips curve and long run Phillips curve.

physical capital

See capital.

portability

Refers to the ease with which money can be carried from place to place.

portfolio investment

Investment in the financial assets of another country. For example, the purchase of Brazilian treasury bonds by U.S. investors would be considered portfolio investment.

positive externality

A spillover benefit of production in which a third party benefits from the production and sale of a good or service without paying for the benefit. An example of a positive externality is the benefit others receive when a person is inoculated against a disease.

preferred stock

A type of stock in which the owner does not have voting rights in the company but is preferred in receiving dividends before common stockholders. In the financial crisis, some firms tried to convert their bonds into preferred stock in order to adjust their balance sheet.

price

The monetary amount that consumers and producers buy and sell some quantity of a good or service.

price ceiling

A legal maximum price. If the price ceiling is below the market equilibrium price, then it is considered effective and results in a shortage of the good or service.

price discrimination

The process by which producers sell the same good or service to different consumers at different prices. Legal forms of price discrimination exist. For example, senior citizen discounts are a form of price discrimination.

price floor

A legal minimum price. A price floor is considered effective if it is above the market equilibrium price and results in a surplus of the good or service. The minimum wage is an example of a government-mandated price floor.

price leadership

Price leadership is a behavior that occurs in oligopoly where one firm usually takes the lead in making changes to prices in the industry.

price war

A condition in oligopoly where firms each try to undercut the others' price. After the price war has ended, the industry usually returns to the price leadership mode.

private sector

Households and firms in the circular flow model are collectively referred to as the private sector.

producer price index

A measure of producer price inflation that acts as a leading indicator of future consumer price inflation.

product differentiation

An economic behavior in which firms attempt to make their product unique from the competition by either making real changes or through advertising. Product differentiation, if successful, gives firms limited pricing power and results in economic profits.

production function

The production function is a relationship that shows how the amount of a firm's output changes as the firm varies the number of a single input that it employs. For example, a production function might look at the variation in the number of hamburgers made at a fast-food restaurant when the firm employs different amounts of labor.

productivity

The amount of output produced with a given amount of resources. As productivity increases, the unit cost of production decreases.

profit

Revenue in excess of costs.

progressive tax

A tax that takes a larger portion of income from high incomes than from lower incomes. The federal income tax is an example.

proportional tax

A proportional tax is one in which all tax payers pay the same percentage of income. A flat tax is an example of a proportional tax.

public good

A public good is one that is nonrival and nonexcludable. The private sector has little incentive to produce these types of goods, so the taxsupported public sector provides them. Examples of public goods include roads, bridges, police protection, national defense, and public schools.

public sector

All levels of government, from local to federal. The public sector trades with the private sector and foreign sector in the circular flow model.

purchasing power parity

See law of one price.

pure monopoly

A market dominated by a single seller.

quantitative easing

A term coined in the 2007–2008 financial crisis that refers to the extraordinary measures taken by the Fed to ensure that banks and other financial institutions remained solvent and able to lend. The hallmark of quantitative easing was the fed purchase of financial assets other than U.S. treasury securities.

quotas

Limits on trade.

real GDP

The gross domestic product measured in constant dollars so that inflation does not affect year-to-year comparisons.

real income

Consumers' income adjusted for the effects of inflation.

real interest rate

The interest rate that equates the level of saving and borrowing in the absence of inflation.

real investment

The author's term used to emphasize the difference between financial investment and investment in physical capital. Real investment involves borrowing to invest in tools, machines, equipment, factories, and new construction.

Recovery Act of 2009

The most recent example of expansionary fiscal policy aimed at stimulating the economy.

regressive tax

A tax that collects a greater portion of income from low-income earners than from high-income earners. Sales taxes are an example of a regressive tax.

remittances

Transfers of money from individuals in one country to individuals in another country not based on trade or investment. For example, when immigrants from Mexico work in the United States and then send their money back to Mexico, they are making a remittance.

rent

The payment for land is referred to as rent.

representative money

Paper currency redeemable in some commodity, usually a precious metal.

repurchase agreement

Also known as a repo, in a repo transaction, the Fed lends money to the dealers, sometimes overnight, in exchange for a treasury security or other high-quality financial asset. This has the effect of temporarily increasing the amount of available excess reserves in the banking system

required reserve ratio

This is the ratio of demand deposits that a bank may not lend against. The required reserve ratio ensures that banks have the necessary cash to meet the transaction needs of their customers.

required reserves

The reserves that a bank may not lend. The required reserves are determined by the required reserve ratio established by the Fed.

reserve requirement

See required reserve ratio.

reserves

The sum of required and excess reserves in a bank. Reserves are an asset balanced by deposit liabilities and shareholder's equity.

revenue

The income of a firm. Revenue is determined by multiplying the quantity of goods sold by the price at which they were sold.

reverse repurchase agreement

If the Fed wants to temporarily reduce the amount of excess reserves in the system, they engage in reverse repurchase agreements, or reverse repos, with the dealers. In a reverse repo, the Fed borrows cash from the dealers in exchange for treasury securities on a short-term basis. This temporarily decreases the available excess reserves from the system.

rival good

A good is considered rival if one person's consumption of the good prevents another person's consumption of the good. For example, a Twix bar is a rival good. If you eat it, then your friend cannot.

rule of law

The idea that government officials must obey the law and that none are above it. Rule of law is conducive to economic growth.

Say's law

The idea that supply creates its own demand. Say's law is a theoretical underpinning of the classical contention that the economy tends to remain fully employed without government intervention.

scarcity

The universal condition that exists because there is not enough time, money, or stuff to satisfy everyone's needs or wants.

securitization

The process by which mortgage loans were packaged together and sold as a type of bond. Securitization allowed banks to make many more loans but created a moral hazard as lending institutions did not bear the risk if the borrower defaulted.

shoe-leather costs

A criticism of inflation that says increases in inflation also increase the number of necessary transactions as money quickly loses value.

short run

In microeconomics, the short run refers to the period of time in which at least one input to production remains fixed. In macroeconomics, the short run refers to the period of time in which input prices do not adjust to changes in the price level.

short-run aggregate supply (SRAS)

Assuming that input prices do not adjust to changes in the price level, increases in the price level provide firms with increased profits and the incentive to increase output. As a result, in the short run, inflation leads producers to increase output.

short-run equilibrium

Any intersection between short-run aggregate supply and aggregate demand that does not occur at full employment.

short-run

Phillips curve A short-run trade-off between inflation and unemployment that assumes that people's inflationary expectations remain unchanged.

shortage

The condition that results when the price of a good or service is lower than the market equilibrium price. A shortage is characterized by the quantity demanded exceeding the quantity supplied.

socialism

An economic system in which government owns and operates many of the factors of production, but households are able to privately own property.

socially optimal

The condition that is met when the marginal social benefit equals the marginal social cost. When externalities exist, the market outcome and the socially optimal outcome are different.

stability

Stability exists when money's value does not vary too much.

stagflation

The economic condition in which both inflation and unemployment simultaneously increase. Stagflation is caused by supply shocks that result in higher input prices for firms.

standard of value

One of the functions of money. When money acts as a standard of value or unit of account, it is being used to measure the relative value of a good or service. For example, prices are stated in monetary terms, so money is being used as a standard of value.

stock market

The market where shares of stock in corporations are bought and sold. Most transactions in the stock market are for shares that have been previously owned by other investors and do not necessarily indicate a flow of savings to firms.

stock option

A stock option is a derivative of stock that allows the owner to purchase shares of stock at a predetermined price.

stockholder equity

This is the ownership interest that the shareholders have in a firm. On a balance sheet, stockholder equity equals assets minus liabilities, and it is similar to the concept of net worth.

stocks

A financial investment that allows the purchaser to own part of a corporation. As an owner, the stockholder is entitled to a share of the corporation's profits.

store of value

Money acts as a store of value when you get it today and are still able to use it later.

structural unemployment

Unemployment that is caused by the permanent destruction of jobs in a dying industry, a mismatch between the skills necessary for employment and the seekers' skill sets, and government programs that create incentives to remain unemployed.

subsidies

Payments from government to the producers of certain commodities, goods, or services. Subsidies create an incentive for firms to produce more output than would otherwise occur in a competitive market. Producer subsidies increase supply, while consumer subsidies increase demand.

substitutes

Goods are substitutes if the cross-price elasticity is greater than zero. For example, when the price of gasoline increases, the demand for ethanol (a substitute fuel) increases.

substitution effect

One of the reasons for the law of demand. The substitution effect occurs when the price of one good changes relative to the price of a related good and consumers substitute away from the higher-priced good to the lower-priced good.

supply

The direct relationship between price and quantity because of producers' willingness and ability to sell or produce output at the various prices that occur in the market.

supply-side economics

A school of economic thought that focuses on providing incentives for producers to increase output by lowering taxes and deregulating industry. Supply-side economics was made popular during the Reagan era, and today many conservative politicians still make supply-side arguments when proposing policy.

surplus

The condition that exists when the quantity supplied exceeds the quantity demanded. Surpluses result from prices greater than the market equilibrium price.

tacit collusion

When firms are unable to legally collude, tacit collusion might result from repeatedly following a tit for tat pricing strategy.

tariff

A tariff is a tax on trade.

taxation

The process by which government coerces revenue from households, firms, and the foreign sector.

Tea Party

A relatively new political movement that promotes low taxes and reduced government influence in the economy. Many of the Tea Party's ideas are similar to supply-side economic policies.

technological monopoly

A technological monopoly occurs when a business is able to exercise a patent or copyright in order to be the sole producer of a good or service.

term auction facility

In response to banks' unwillingness to borrow directly from the Fed's discount window, the Fed instituted the term auction facility, or TAF, which auctioned off blocks of short-term loans to banks. The purpose of the TAF was to ensure that banks had the liquidity necessary to continue their lending activities.

The Great Recession

A term used in the media to describe the recession that lasted from 2007–2009.

total cost

Fixed cost + variable cost.

trade

The act of exchanging goods and services.

traditional economy

In a traditional economic system, the questions of what and how to produce, and who to produce for, are answered by tradition or custom. Indigenous groups from around the world are just a sample of the people who have practiced or continue to practice traditional economy.

tragedy of the commons

The idea that when a resource is held in common, then the individual users of the resource have no individual incentive to conserve the resource.

tranche

Part of a whole. The term is used in describing the different loans of various risk in CDOs.

treasury bill

Treasury security with a maturity of less than one year. Treasury bills or T-bills provide investors with a liquid asset while providing the government with cash.

Treasury bond

Long-term treasury securities are referred to as treasury bonds or T-bonds. Treasury bonds mature after 30 years.

treasury note

Medium-term treasury securities with maturities of two to ten years.

Troubled Asset Relief Plan (TARP)

The program enacted at the end of 2008 that gave the treasury secretary the power to step in and help recapitalize banks that had been hit by the defaults in the CDO market.

U1

The unemployment rate that only includes people unemployed fifteen weeks or longer, as published by the BLS.

U2

The unemployment rate that only includes people who have lost a job as opposed to those who have quit or those who have entered or reentered the labor force, as published by the BLS.

U3

The official unemployment rate published by the BLS.

U4

The unemployment rate that adds discouraged workers to the official unemployment rate, as published by the BLS.

U5

The unemployment rate that includes all marginally attached workers, as published by the BLS.

U6

The most all-inclusive measure of unemployment as published by the BLS. Includes all those listed in U1–U5 plus those who are employed part time because of economic reasons.

unemployed

The condition of not having a job but being a member of the labor force. To be considered unemployed, a person must be jobless yet actively searching for a job by having searched in the last four weeks.

unemployment rate

The unemployment rate is equal to the number of unemployed persons divided by the number of people in the labor force.

usury

Traditionally, usury was considered the charging of interest on borrowed money. Today, usury is understood to mean charging excessive interest on borrowed money.

utility

The amount of usefulness or satisfaction a person gets from performing an activity.

utility maximization

Where firms try to maximize profits, consumers and individuals attempt to maximize utility. This can be understood to mean that individuals try to maximize usefulness or personal satisfaction.

utils

The amount of utility or happiness you get from doing something.

variable cost

Costs that change as a firm's level of production changes. Examples of variable costs include hourly wages, raw materials, utilities, and per unit taxes.

vertical merger

A business combination between firms in the same industry but at different stages of production. For example, Andrew Carnegie reduced the cost of steel production and dominated the steel industry by vertically integrating. Carnegie Steel controlled production from iron ore to finished products.

voodoo economics

See supply-side economics.

wage-price spiral

A continuous cycle of inflation that is set off by workers demanding higher wages to compensate for higher prices. The higher wages result in even higher prices, and the process then repeats.

wealth

The sum value of all you own.

World Trade Organization (WTO)

The World Trade Organization is an international body of countries committed to reducing trade barriers and promoting free international trade. The WTO is the successor to the General Agreement on Tariffs and Trade, or GATT.

zero bound

The limit to traditional monetary policy. Nominal interest rates cannot be reduced below 0%. For most of 2009 and 2010, the Federal Reserve maintained a target for the federal funds rate of 0%–0.25%, effectively reaching the zero bound. In order to encourage investment and consumption, the Fed engaged in quantitative easing in addition to maintaining low interest rates. See liquidity trap and quantitative easing.

zero-coupon bond

A bond sold at a discount from face value. Treasury notes are examples of zero-coupon bonds. To compute the interest rate, subtract the purchase price from the face value and divide by the purchase price.

Printed in Great Britain
by Amazon